I0020125

AI in Human Terms

David Lloyd

© David Lloyd, 2024

All rights reserved. The use of any part of this publication reproduced, transmitted in any form or by any means, electronic, mechanical, for the copying, recording, or otherwise, or stored in a retrieval system, machine learning model, search system, without the prior written consent of the publisher or, in the case of photocopying or other reprographic copying, a license from the Canadian copyright licensing agency is an infringement of the copyright law.

Library and Archives Canada Cataloging in
Publication data is available upon request.

ISBN:
978-1-7383731-1-6 (Hardcover)
978-1-7383731-0-9 (Paperback)
978-1-7383731-2-3 (eBook)

Cover Illustration © O-IAHI, Shutterstock, Licensed

Cover Design © David Lloyd

Printed and bound in Canada

www.aiinhumanterms.com

For my family who have always supported me and my crazy ideas, with love. To my wife Kim, providing me with unwavering support and guidance I should listen to more often, my daughter Samantha, for your drive to succeed, love of commas, and gelato, and my son Erik, for your resilience, passion, and occasional Call of Duty™ breaks.

-finally-

To Darren Redfern (1966 – 2022) who could take the most complex topics and help everyone in the room understand them.

Table of Contents

Intro: How Did We Get Here So Fast?

Imagine waking up one day to find that artificial intelligence, once a distant dream, is now intricately woven into the fabric of your daily life.

In recent years, as I've spoken with a diverse group of individuals from parents, businesspeople, retail workers, to medical staff and many managers across industries - it's become clear that artificial intelligence, or AI, has taken many by surprise. The common refrain during these conversations is one of disbelief: "This came out of nowhere. What happened?" Yet, as sudden as the advent of Generative AI (GenAI) might seem, its roots are deep and complex, often overlooked in our daily hustle.

1950's	1960's	1970's	1980's	1990's	2000's	2010's	2020's
AI Founded 1946	Term "AI" Coined 1956	First PC 1977		WWW 1993		IPhone 2007	Data & Compute 2017	GenAI 2022

Figure 1 - Timeline

Let's start with a brief timeline, a favorite starting point in my presentations. The journey to today's AI began decades ago, not with flashy startups or tech juggernauts, but with the bulky mainframes of the mid-20th century. These evolved into minicomputers and, eventually, into the personal computers we know today—like the Apple II launched in 1977, a landmark in computing history.

Following personal computers, the next significant leap was the internet, introduced as the World Wide Web in 1993. Originating from

Advanced Research Projects Agency ARPANET in the late '60s, it connected researchers globally and, after years of development and the convergence of key technologies like browsers and web servers, gained massive traction by the late '90s.

Up next, mobile technology, too, transformed our lives. Recall the early car phones of the late '80s—a precursor to the smartphones that would redefine our communication landscape with the introduction of the iPhone in 2007. These devices merged touch screens, internet connectivity, and apps into an indispensable tool.

Fast forward to November 2022: the launch of ChatGPT by OpenAI. Suddenly, AI was not just a utility but a household name, with roughly 100 million people interacting weekly with ChatGPT and 1.6 billion visits to the OpenAI site over a month. To grasp the rapid emergence of GenAI, we must look back to foundational technologies like Google Translate and other natural language solutions that set the stage years earlier, as well as non-generative AI approaches.

These technological shifts didn't just appear; they evolved, unnoticed, becoming as essential as the internet, laptops, and mobile phones are today. AI has been a quiet companion in our technological journey, tracing back to pioneers like Alan Turing in 1946 and computing in general with Ada Lovelace, whose 1843 "Notes" contained what is considered the first algorithm.

Let's delve into the impact of AI understanding the concepts and history followed by a focus on Large Language Models (think ChatGPT). The goal is to demystify these technologies using relatable terms, making clear how integral AI has become to our modern existence.

David Lloyd

Scraps of Evolution

"...the kind of control you're attempting is not possible. If there is one thing the history of evolution has taught us, it's that life will not be contained. Life breaks free, it expands to new territories. It crashes through barriers. Painfully, maybe even... dangerously, but and...well, there it is."

--- Malcolm to John Hammond, Jurassic Park[1]

If there's a lesson to be learned from *Jurassic Park*, it's that our creations often take on lives of their own. Transform ancient mosquitoes into dinosaurs, and those very dinosaurs might just fight back. View life—or, for our purposes, technology and specifically AI—from any angle, and it inevitably swerves from our most meticulous plans. Reflect on this iconic line from *Jurassic Park*, substituting 'life' with 'AI':

There's a poignant reason I chose *Jurassic Park* to start a dialogue about AI. Imagine if the events of Michael Crichton's film unfolded today—dinosaurs roaming through New York City, theme parks brimming with visitors, and an investor frenzy around the latest Elon Musk venture, 'TyrannosaurusX'. When we substitute 'life' with 'AI' in the dialogue, the analogy sharpens: in recent years, AI has been 'breaking free' on its own terms.

---

[1] Spielberg, Steven, dir. *Jurassic Park*. Universal Pictures, 1993. Based on the novel by Michael Crichton, screenwriter, Michael Crichton and David Koepp

Technology sneaks up on us. Suddenly, OpenAI's ChatGPT bursts onto the scene, and within months, what seemed like overnight, its user base exploded to over 150 million by May 2023. ChatGPT swiftly became a term as common as household names. The entire world buzzed about artificial intelligence—a concept so potent it captured global imaginations.

Yet, AI isn't a sudden phenomenon. But if it wasn't for putting a friendlier face on AI, democratizing it's use, we may have been waiting longer.

For nearly seven decades, artificial intelligence has been brewing, growing significantly more sophisticated over the past twenty years. AI didn't just appear; we nurtured it, fed it with our data, our daily technology use—connecting with friends, deciding what to watch, ordering food, navigating cities.

We enabled its evolution.

Now, we face an urgent question: How will AI 'find a way'? Humanity inches ever closer to crafting an intelligence that could surpass its creators. AI's potential to enhance or threaten our way of life hangs in balance. As artificial intelligence seeps into consumer hands worldwide, we hit a pivotal moment. It could surpass us, and stringent regulations may be necessary and unable to curb the unchecked ambitions of AI firms. Understanding AI's roots—and its trajectory—is critical.

We need to drill into the amber, extracting the DNA that shaped today's AI. Understanding how to teach a computer to complete 'I want peanut butter and...' with 'jelly' or use a Shakespearean twist to create 'a spread of peanut paste' requires no math or computer science degree, just curiosity.

For almost seven decades, we've been unraveling the artificial brain—pondering whether a computer can play chess, solve puzzles, or write essays. Today, AI's capabilities challenge the breadth of human knowledge, raising profound questions:

Should we, or shouldn't we?

As we stand at this critical juncture, our choices will shape the future. AI demands our engagement. Deciding whether AI will redefine our work and life, knowledge—particularly human knowledge—is our greatest ally. We need to deconstruct the complexity of AI's concepts and history to navigate our future.

We are not mere bystanders in the saga of AI. We are its architects and benefactors. The power that AI affords us is unprecedented—we can communicate across languages we don't speak, predict calamities, and save lives, yet we also face risks like eroding privacy and escalating surveillance.

The choice is ours. Being informed is essential. As AI's DNA continues to weave into our daily existence, albeit often unnoticed, becoming part of our lives. How we choose to interact with it, what we expect from others, corporations and governments, and how we harness its potentials or mitigate its risks, will define our future.

This book is structured in three parts: a breakdown and explanation of key AI concepts simplifying the jargon (in human terms), a concise history of AI, and a brief spotlight on large language models powering generative AI. While you can explore each part as you'd like I invite you to join me on this AI journey— through its theoretical underpinnings, it's origins, to understanding Generative AI. Who knows? We might even encounter some proverbial dinosaurs along the way.

Part 1: Acronyms, Math, and Science, Oh My

"Pure mathematics is, in its way, the poetry of logical ideas."

--- *Albert Einstein*

While the realms of math (from which AI primarily springs) and science (specifically computer science and machine learning) can seem as daunting as a Jurassic T-rex, there's no need for panic or a frantic escape in a Jeep. Our journey into AI won't require deep scientific knowledge or familiarity with endless three-letter acronyms that might send you checking your rearview mirror in terror. Instead, we're here to demystify AI, breaking it down into accessible, practical terms—no dinosaurs included.

This part attempts to strip away the complexities of AI, but introduce some specific terms. You don't need to be a science, math, or philosophy major; just bring your sense of adventure and perhaps a bit of common sense. Our goal is to explore the multifaceted ways AI influences everyday life without getting bogged down in the theoretic details that researchers and scholars might appreciate. We're not here to build our own AI models; rather, we're here to understand AI in human terms. Behind every daunting concept in science or math lies a basic idea that most people can grasp. Here, we distill AI to its core elements to make the concepts more relatable and less intimidating.

Technology moves at a breakneck pace, with yesterday's breakthroughs quickly becoming today's old news. The use of AI, while a staple at tech giants like Google, Amazon, Meta, and Microsoft for decades, has only recently entered the public consciousness through the advancements in Large Language Models. This is fortunate because it gives the rest of us time to catch up. As we begin to understand these models and approaches, we're better equipped to engage with critical questions: "Should we?" "How will we?" and "What impacts will this have?" In the next set of chapters certain words will be highlighted in **bold** as they represent some of the core terms used in AI. Let's dive in, ready to untangle the complexities of AI and explore its profound implications on our world.

What is AI?

In human terms, think of AI as an 'intelligent machine.' While various definitions exist, at its core, AI is about machines performing tasks typically done by humans, predominantly making predictions. Will you buy a new dress? What's tomorrow's weather? Will a stock price rise or fall? Is that a cat or a dog in the picture? Draft an email for a job application or even paint a picture of your favorite vacation spot in the style of Monet.

These tasks form the basic ingredients of our AI cake. Seen in reverse, these predictions transform into a formula or a model, essentially a mathematical representation that learns from data to predict outcomes. For instance, it can determine if your credit score qualifies you for a credit card or assess the likelihood that you may want to purchase sandals versus a parka.

Models can be dynamic, learning and adapting over time, or static, unchanging unless new data is fed into the system. When someone mentions 'AI,' it's akin to saying, "I like cake." Just as there are many types of cake, there are numerous concepts within AI.

Let's slice this layer cake to better understand it.

David Lloyd

A model is essentially a computer program that continuously learns from data, thus training itself to make predictions—this is machine learning.

Machine Learning (ML) is the bedrock of AI. It includes both supervised and unsupervised learning. In **supervised** learning, we provide the model with both data and labels during training, like predicting a house's price from its square footage. The model learns from the data to predict prices more accurately over time. In **unsupervised** learning, the data lacks labels. Here, the model attempts to organize data into groups by identifying patterns, although the relationship between input and output isn't explicitly known because it does not know what the input and output are called.

Deep Learning, a subset of machine learning, utilizes **neural networks**. What makes it 'deep' are its layers. More layers mean more complexity, deeper thinking, and enhanced accuracy. Deep learning applications range from vision and robotics to language processing and recommendation systems.

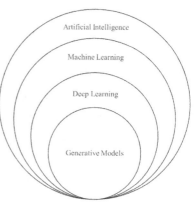

Figure 2 - AI Layers

Other concepts include specialized types of deep learning and **Reinforcement Learning** (with Human Feedback) or RLHF. If you read the second section, the brief history of AI, many of these concepts are introduced there. By breaking down these concepts to their essence, they become easier to understand, and everyday examples can help demystify them even further. So let's break it down.

From Being Told, to Learning

Different areas of AI are interdependent, much like the various systems in a well-orchestrated symphony. Take early expert systems as an example. These programs operated on a straightforward input-output basis, where users, such as doctors, fed in data—perhaps a set of symptoms—and the system checked these against a series of "hard coded" rules derived from the expert's knowledge.

Consider Dr. Smith, a hypothetical physician who encoded his medical expertise into an early AI system. This system might start with the symptom input of a "sore throat" and then follow a decision tree of yes/no questions—like "are you running a fever?"—to diagnose an ailment, perhaps arriving at strep throat.

These systems, whether you call them decision trees, if/then/else statements, or expert systems, were static (or "hard coded"). They didn't evolve unless someone manually updated or reprogrammed the rules. This is a key point because, as AI evolved, it began to shift from rigid adherence to predefined rules towards learning from data in real-time, a fundamental principle of machine learning.

Machine learning marks a pivotal development in AI's history: systems started to modify their responses based on new information. They weren't just repeating what Dr. Smith knew; they were building upon it. This adaptive capability, where AI continuously improves its predictions by learning from outcomes, represents a significant leap from traditional expert systems.

Let's dive deeper into machine learning, arguably a cornerstone concept within AI. Machine learning involves creating models that learn from data to make predictions. A familiar technique from AI's history, "k-Nearest Neighbor" (k-NN), illustrates this by grouping similar data points—like flights and luggage, or bathing suits and sunscreen. Imagine you've just booked a flight to Florida and are shopping on Amazon for flip-flops. Suddenly, you're recommended sunscreen, goggles, and a new swimsuit. This prediction results from analyzing patterns among millions who made similar purchases, a process based on clustering similar behaviors.

If a model labels the input data as "warm travel" and outputs "bought sunscreen," that's **supervised learning**—we know the labels because we've assigned them, a method known as **classification**. If the system is left to identify natural groupings (**clustering**) without pre-labeled data, that's **unsupervised learning**.

See, there are no terrifying dinosaurs here—just fascinating ways AI integrates into and enhances our understanding of the world. Now, let's delve deeper, starting with some foundational concepts.

Model Machines

"All models are wrong, some are useful."

--- *George Box*

Imagine you are house hunting because your family is growing, and you want a bigger place in your favorite city. There are numerous factors to consider: the number of bedrooms, bathrooms, garage space, lot size, the quality of local schools, commute time, transit availability, postal code, and, crucially, the speed of your internet connection.

Now, think of a "**model**" as your personal prediction assistant. This model can sift through more housing options and evaluate more factors than you could feasibly manage on your own, especially if the area you're considering has hundreds or thousands of potential homes. This model has been trained using data on previously sold homes—learning from their **features** (things that describe the house) and their prices (labeled output). When you provide the specifics of a home you're considering, the model predicts the home's value, helping you determine whether it fits your budget and if you're getting a good deal.

At its core, in this case, a model is essentially a formula that has learned patterns from data. Unlike expert systems that follow specific rules, machine learning models derive their understanding from the data they've been trained on. The key here is that the model can make accurate predictions even about homes it hasn't specifically learned about—it **generalizes** from its training to predict against new, unseen data.

Now that your assistant is set up, it can help you weigh the pros and cons of different homes. Behind the scenes, the model processes a vast amount of data—considering how features like square footage, location, number of bathrooms, postal code, and school district might affect the price you'll end up paying.

Despite the immense complexity of the data, you're presented with a streamlined decision: the model estimates you should expect one home to be worth about $345,000 and $390,000 for another. Interestingly, the first home is listed at $368,000, and the second at $375,000. With thousands of data points at its disposal, the model provides you with a simple prediction to decide which home offers the best price based on its' features. This scenario, where you make decisions based on the model's predictions, exemplifies a concept known as "**human-in-the-loop**".

Whether it's predicting the presence of a car or pedestrian, the price of a home, the next word in a sentence, a language translation, or your evening movie pick, a model is a program trained on data to make predictions. Its effectiveness grows with the amount of data it processes.

However, not all models are created equal. What different approaches exist in building these models? That's what we'll explore next.

Machine Learning

There are a few fundamental concepts to remember as we dive further into artificial intelligence, specifically into machine learning. Models, much like babies, aren't well adapted to the world. They are dependent on parents, teachers and self-exploration to build an understanding of their surroundings. Think of models in this sense (although children are another issue entirely). Models all need to be initially trained to learn, with success being rewarded and mistakes corrected.

Machine learning differs from earlier systems. Here, the goal is not to have to write explicit programs or instructions to accomplish tasks and make predictions. As we discuss later, in the early part of AI history, tools like expert systems were quite prescriptive about the decisions and actions the model would take. However, this approach only works if you wish to continue explicitly telling the program what to do. You will need to update it periodically and continue programming it for all the permutations and changes that may occur over time. It does not learn. It does what you tell it to. In the case of machine learning, the goal is to build a model that can continue learning from any new data it receives making new novel predictions.

Machine learning is about writing as little computer code as possible: you're letting the machine do the work. By studying the data provided, the computer learns the rules or patterns available in the data, then it can make predictions. At its most basic level, even new Generative (the G in GPT) AI tools (GenAI) like ChatGPT, Gemini, and others are simply a model (software program) using prediction based on learning. The fact these tools can accomplish this without explicit programming is proof of how powerful these technologies are becoming.

This is why we are hearing so much today about the quality of data, the nature of bias in data, and

whether the predictions the model is making are
transparent (or explainable).

For example, try asking for the word that follows "peanut butter and...": In most North American households, the answer would be "jelly." In the voice of William Shakespeare, it might be peanut butter and a "nectar of fruit," as the concept of jelly was not invented yet (but we know he loved poetic prose). In Spanish, mantequilla de cacahuate y "mermelada," where mermeladde is Spanish for jelly or jam, depending on the region. No one explicitly programmed those models to do this: they learned it all on their own.

This is an example of where the volume and type of data used in training models can become a challenge. Since these models have been built on ingesting all the digital content they can find (typically the internet), the models are influenced based on the patterns of that data. For example, if you speak a language that has not been digitized or does not have a large amount of digitized data associated with it, then the reinforcement of those words will be far less. If you have a language that is passed through generations by speaking it, but not writing, that language as it relates to machine learning would in essence not form any material part of a large language models learning. Negatively, if these models ingest everything they find digitally, then hate speech, inaccuracies, racism, ethnic and gender biasness are all part of the model if it remains uncorrected. And if statistics and vectors on how words relate continue, then the "louder" data wins.

Before we delve into the intricacies of supervised and unsupervised machine learning, let's consider how pervasive machine learning already is in our daily lives—so much so, you might not even notice it. Take, for instance, email filters. These tools have evolved over the years, becoming increasingly sophisticated in weeding out unwanted emails from our inboxes (though, admittedly, we might wish they could handle spam texts just as effectively). Initially, these filters had to learn from the data fed into them, understanding patterns in words or identifying typical senders of spam. Over time, they became adept at relegating emails from a so-called rich prince in a distant land or a chain letter threatening bad luck to your junk folder. This progression in machine learning capabilities demonstrates not only the technology's growing sophistication but also its integration into the fabric of our everyday digital interactions.

How does this learning occur? Initially, the model acts on your explicit instructions. For instance, if you decide that a chain email warning of a curse belongs in the spam folder, the model notes your preference. But it doesn't stop there. The model can also tap into the wisdom of the crowd. If a multitude of users are consistently marking emails from a supposed rich prince as spam, the model learns to automatically block similar messages.

Additionally, your email provider plays a crucial role by identifying which companies are legitimate senders and which aren't. The model observes patterns in the emails' subjects, senders, and specific keywords such as "prince," "send," and "gift cards." Through these insights, it refines its ability to discern spam from genuine communication. This collaborative learning process, combining direct user feedback with broader behavioral patterns, allows the model to enhance its accuracy and responsiveness over time.

Another set of examples includes movie, book and other recommender systems like the ones in Netflix, Audible, Instagram, or Tik Tok. These systems use both personal and group data, combined with machine learning, to look at past choices and recommend (predict) what you might want to read, watch, or listen to next. They also use reinforcement learning based on how you and others may have rated a particular music artist, book or video, therefore driving different selections in the future. Soon the system might start playing or recommending more/less of certain authors, artists or videos.

Another familiar application is virtual assistants like Siri or Alexa and now a massively growing group of assistants based on using generative AI (large language models), which have evolved from simple task performers to our AI sidekicks. We've come a long way from the early days of "Clippy" the friendly paperclip. These assistants process your commands—whether typed or spoken—and use machine learning to discern your intentions and assist accordingly. Specifically, these applications focus on taking the "utterance" or "**prompt**" (the command you typed or spoke) and use machine learning to determine the intent of what you are trying to do.

More generally, these models determine the grouping of the utterances associated with a certain intent. For example, teaching the machine learning model that "take time off", "go to Mexico for 2 weeks", or "go on vacation" are all the same intent. When the virtual assistant understands the intent, it can assist with the task: asking when you want to start your vacation, or where you may wish to go.

Machine learning also plays a critical role in more specialized fields like medical diagnostics. Here, models trained on vast image libraries can predict medical conditions from new images, aiding in early detection and treatment strategies for cancer and other diseases. In practice, the applications for these types of technology are virtually infinite. Medical devices such as glucose readers and heart monitors can be coupled with machine learning to aid in diagnosis or provide proactive advice (should you watch your sugar intake as your glucose is rising). Are you potentially suffering from atrial fibrillation?

Non-medical applications for machine learning include fire detection. In use in Southern California with local utilities and now in eastern Canada, the use of image analysis for possible fires given the wildfire conditions enables an early warning. Video content is analyzed for potential fire indicators resulting in notification to nearby fire departments or fire watchers to validate the potential start of a wild fire. The human-in-the-loop helps in the verification process with a system that saves lives and potentially hundreds of millions of dollars in costs to re-build. It also helps further train the model for nuances in the images for fire detection.

The past 5 years have seen a continual expansion of both combustion and electric vehicles containing machine learning within their array of sophisticated technologies: LIDAR, RADAR, and even some of the technical gadgetry that "Shaky" the robot (covered in the history section), contained during the early days of robotics.

Machine learning is everywhere. What is especially interesting is how these different technologies have been brought together. Combined, they create a massive stream of data capable of understanding something as complex as driving in an urban environment. Here, the use of machine learning has aided in developing accident avoidance: it knows where the vehicle is on the road, and where the other vehicles are surrounding it. This ends in a much safer driving environment, especially once most vehicles are using the technology. It does not remove the human element or driver, we are still the human-in-the-loop but uses many different machine learning techniques to account for possible human error. Other examples of machine learning in use today include predicting weather, translating languages, and the stock market (although if someone has truly nailed that prediction capability, they aren't sharing it).

In summary, machine learning teaches a computer to become smarter over time. Like raising a child, the computer's comprehension grows as it is provided with examples that are labeled (supervised) and at times letting it discover examples itself with unlabeled data (unsupervised). When the model gets the answer right or wrong, techniques like **Reinforcement Learning with Human Feedback** (RLHF) can be used to rate a response or simply click on a thumbs up or down. It's not much different than how a baby learns, where the reinforcement might be the smile on a parent's face or supportive words. For those more accustomed to dogs and cats, it's teaching a pet new tricks through rewards and repetition (although my cats might make me question that).

The point is computers can learn. This learning has applications far beyond what you or I have considered – and this is only the beginning. Now that we understand the general concept of machine learning, it's time to dive into some of the details.

Training Models: Supervised and Unsupervised Learning

"The brain is showing you 'we're excited, we want it we'll follow it."

[he guides dogs to the platform]

"Then you come to the point of reference. They're gonna know that once they touch that, they're gonna receive food. [They say] okay, so how do we receive food? We go to the point of reference."

--- Cesar Milan, The Dog Whisperer

Supervised Learning

As we have talked about previously, supervised learning is a type of machine learning where we know the input to the model and have labelled the expected output. Let's consider a real example of how this can be applied in medicine.

Imagine you're an ophthalmologist in a country with a dire shortage of specialists yet tens of thousands of potential patients. In this region, a prevalent issue is diabetes, which often leads to diabetic retinopathy—one of the leading causes of blindness worldwide.

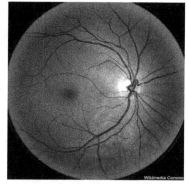

Figure 3 - Retinopathy Image

In the realm of AI, this scenario is addressed through a supervised learning approach. Here, a model is trained on a vast dataset, in this case, images of the retinal fundus—the interior surface of the eye, including the retina. These images are meticulously annotated by professionals to show the presence and severity of the disease. This method was notably implemented by Google in 2016, in collaboration with India's Aravind Eye Care System.[2]

After training and testing, these models can identify diabetic retinopathy across various contexts. In a country of over 1.3 billion people, where there are insufficient board-certified ophthalmologists and a stark disparity in patient distribution, particularly in rural areas, this technology is invaluable. With today's mobile technology, anyone can take a high-resolution image of their eye using a smartphone and submit it for an initial analysis. The aim is early detection—identifying patients before symptoms manifest and ensuring that ophthalmologists focus on patients needing immediate care.

This process exemplifies "supervised learning," where the model is taught to differentiate between successful and unsuccessful outcomes based on the input data it analyzes. Given its exposure to potentially millions of examples, the model's proficiency in making accurate predictions independently improves significantly over time. Thus, even when presented with a retina it has never seen, the model can predict with high accuracy whether it exhibits signs

---

[2] https://about.google/stories/seeingpotential/

of retinopathy. This capability not only demonstrates the power of machine learning but also how it can be harnessed to address critical health care challenges effectively.

Let's explore more general instances of supervised learning in action:

Image Recognition: This application of supervised learning is familiar to anyone who uses social media. When your photo app identifies and suggests names for the people in your pictures, it's using supervised learning. It relies on previously tagged photos as input—both from your own collection and possibly from others. The output is the suggested tags. Essentially, the app has 'learned a face' and can now suggest who might be in a photo based on previous tags and examples it has encountered. However, when the model encounters a face, it doesn't recognize, it asks you to identify the person. Once you provide a name, the model links this new input with the face, enhancing its ability to recognize that person in future photos.

Recommendations: Platforms like Netflix, Amazon, TikTok, Instagram, and Spotify use supervised learning to tailor recommendations. These models analyze a blend of content you've engaged with, items you've saved to your "wish list," and content liked by others who have similar tastes. The model builds a profile based on this data, which includes your personal information like age and gender, as well as your interaction history. This profile helps predict what you might enjoy next, suggesting new content in similar genres or by similar creators. The recommendation system continuously refines its predictions based on your ongoing interactions and the evolving preferences of users like you.

These examples illustrate how supervised learning can enhance our daily digital experiences, from keeping our memories organized to helping us discover our next favorite song or show. Each interaction we have with these technologies' feeds into a cycle of learning and improvement, making the models more accurate and effective over time.

Unsupervised Learning

If supervised learning relies on defined inputs and outputs to train a model, then unsupervised learning allows the model to make discoveries on its own. Imagine a young child given a box of colored blocks, not yet familiar with the names of colors or the shapes like triangles, circles, or squares. Spread out on the floor, the child begins to organize these blocks

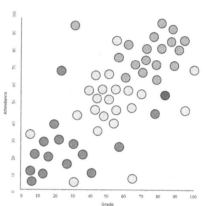

Figure 4 - K-Means Cluster Example

by color or shape—blue triangles here, orange squares there—without knowing what these represent, only that the items share characteristics.

This is akin to how unsupervised learning works in AI. Instead of feeding the model explicitly labeled examples, we give it a vast array of data and let it find patterns or groupings by itself (clustering). Later, we might choose to label these groupings if needed (classification). It's like sorting blocks without prior knowledge of what they signify. Let's examine how this plays out in a marketing context, a scenario most of us have likely contributed to through online shopping.

Take an example in online retail. Imagine you're an online fashion retailer curious about your customer base but lacking detailed information about their characteristics. However, you have accumulated substantial data on individuals who have made purchases. These features might include age, gender, postal code, cart size, purchase frequency, and other relevant data, but the model does not know these labels. Here, for example, you might employ a method called **K-Means Clustering**, which we briefly touched upon earlier. This technique will assess the various data features and begin to group individuals with similar characteristics without initially understanding or labeling these clusters.

Through the lens of unsupervised learning, you might discover distinct groups within your customer base, which you could eventually label as "budget-conscious," "premium customers," "summer shoppers," or "families with children." Understanding these clusters allows you to tailor marketing strategies, such as targeted email offers or personalized shopping experiences, aiming to enhance purchase rates by catering specifically to revealed preferences.

Other Applications

Meal Delivery: Take data from a meal delivery service, for example. An unsupervised model might analyze all the meals ordered and group them by the type of food—though it doesn't initially recognize these as food categories. The chef might discover that there are significant clusters of meals that are vegetarian or seafood preferences, leading to adjustments in the menu to cater more to these tastes, thus enhancing customer satisfaction and potentially increasing business.

Recommender Systems: Music streaming services like Spotify analyze millions of user playlists. They might notice that fans of certain Taylor Swift songs might enjoy certain ones by The Foo Fighters. Even though these artists don't share a genre, the model creates clusters based on user preferences, enabling it to recommend songs that, while seemingly unrelated, are likely to appeal to the listener's tastes.

<u>Social Media Content</u>: A model could examine thousands of social media posts and group them based on intent, tone, sentiment and other similarities. Without initially knowing the underlying topics, it can still match these patterns to content you usually engage with and suggest new posts that might interest you.

In essence, unsupervised learning lets computers uncover hidden patterns in data without direct guidance. It's akin to solving a jigsaw puzzle without seeing the picture on the box. As the pieces—data points—come together, the model forms a coherent image on its own. Once refined with labels, this unsupervised learning can transition into a supervised approach, ready to tackle more defined tasks with a richer understanding of the data landscape.

Reinforcement Learning

> "A strange game. The only winning move is not to play. How about a nice game of chess?"
>
> --- WOPR/Joshua, Wargames[3]

Pets, particularly puppies, provide an excellent analogy for understanding the mechanics of **Reinforcement Learning** (RL). Consider teaching a puppy to sit. Initially, the puppy doesn't grasp what "sit" means. You might gently press its hindquarters to encourage the sitting motion, rewarding it with a treat for each successful attempt, and withholding rewards when it strays from the command. Over time, the puppy learns to associate sitting with getting a treat. This fundamental process—rewarding closer approximations to the desired behavior until the correct behavior is consistently achieved—is precisely how RL models operate in the

---

[3] Wargames, 1983, directed by John Badham, written by Lawrence Lasker, Walter Parkes and Walon Green

development of AI. Weaning your puppy off treats is another issue altogether.

In RL, models make predictions, and these predictions are reinforced with rewards or penalties. For example, a model might receive a "thumbs up" for a correct prediction or a "thumbs down" for an incorrect one or perhaps 5 stars versus 1. If this example is done by a person, this is reinforcement learning through human feedback (RLHF). These signals guide the model, helping it to make better choices in similar future scenarios. As the model iterates through trials, it learns from its experiences, refining its strategies to maximize rewards.

Let's consider some practical applications:

Games: Reflecting on the early use of AI in checkers, we see that reinforcement learning allowed the program to learn optimal moves against human or AI opponents. This technique was also used in more complex games like AlphaGo. As the game progresses, each move is evaluated and the successful strategies are reinforced, making it likely that the model will choose similar moves in comparable situations in the future. Interestingly to train these models faster, models were created that trained and competed against each other. In this case they never grew tired and provided a method to reinforce learning with model-to-model interaction.

Robotics: Shakey, an early robotic platform, shares similarities with modern devices like the Roomba. As the Roomba navigates a room, it learns to avoid obstacles through trial and error—much like Shakey navigating around baseboards. Other newer versions of Roomba have additional sensors that enable them to visualize obstacles, not simply bump into them to remember. This reinforcement through direct interaction with the environment is a key element of how these robots learn to function more effectively over time.

Social Media: Platforms such as X, Meta (Facebook, Instagram, and Whatsapp), and TikTok incorporate reinforcement learning in how they curate and recommend content. User interactions, such as likes, dislikes, and viewing duration, provide feedback that helps these platforms refine their algorithms to better predict and serve content that engages users.

Furthermore, reinforcement learning extends to the fine-tuning of Large Language Models (LLMs) discussed in Part 3. For example, a model might generate multiple responses to a prompt, and a human reviewer selects the response that best matches the desired conversational tone. This feedback loop gradually teaches the model to produce responses that are increasingly aligned with human preferences, enhancing the effectiveness of applications like chatbots or digital assistants.

In summary, reinforcement learning is a dynamic method whereby models learn optimal behaviors through continuous feedback. It's a process that never tires, allowing for constant refinement and adaptation without the limitations of human fatigue or boredom. This makes RL an incredibly powerful tool in the development of AI across various applications—from gaming and robotics to social media and beyond.

Neural Networks

"Ogres are like onions, they have layers."

--- Shrek

In our discussion of AI history in Part 2 we will talk about mazes and minds. These concepts still resonate in modern AI approaches. Some forms of machine learning, like reinforcement learning, use simple symbols—such as a thumbs up or thumbs down—to curate content likely to engage us, akin to navigating a maze. On the flip side, neural networks, which utilize billions of **parameters** (even trillions in more recent models), act as simulations of the mind, attempting to replicate the intricate operations of the human brain.

In these neural networks, the simplest units are known as perceptrons functioning similarly to neurons. They are far simpler than their biological counterparts but serve a similar purpose: to process and pass on signals. In the human brain, synapses connect neurons and modulate the strength of signals. In neural networks, connections between perceptrons (found in the layers, as we will explore), aim to mimic this variability, adapting and learning much like our own brains do in response to new information. These are more precisely called deep neural networks due to hidden layers we will discuss shortly.

As we delve deeper into AI's capabilities, we see our own cognitive processes digitized, highlighting the profound potential of these technologies to mimic—and sometimes enhance—human thought and interaction.

Deep Learning

Imagine you're preparing for a birthday party, assembling snacks like 7-layer dip, lasagna, and kettle chips, but you've forgotten the cake. Luckily, you have a neural network with deep learning capabilities at your disposal to help you whip up a chocolatey layered treat. But how does a neural network relate to baking a cake?

Consider the elements involved in the cake: the ingredients (flour, sugar, cocoa powder, eggs, vanilla, milk, etc.) are your inputs. The mixing and baking process involves several hidden layers—transformations that turn these simple ingredients into a delectable cake. This analogy approximates how a neural network functions, with inputs processed through multiple layers to produce an output—in this case, a cake. You possess all the inputs you'd need. Now, you are going to use various amounts of each ingredient to create it, those are **weights**. The weights could, in this example, also be the amount of ingredients, mixing time, oven temperature, and cooking time. You will notice in the figures that these parameters (grey balls) and weights (lines between them) are all interconnected, much like the neurons and synapses mentioned previously, sending signals to each layer.

Input Layer

Figure 5 - Deep Learning (Input Layer)

As we will discuss in Part 2, Yann LeCun introduced us to the concept that deep learning is 'deep' because it incorporates multiple layers. Think of your deep neural network like making a cake, where each layer of the process and the ingredients (inputs and parameters) represent tweaks and changes you make to perfect your recipe.

Each neuron in a neural network's hidden layer processes inputs from the preceding layer, combining them in complex ways to mimic the nuanced actions you might take while mixing ingredients. For instance, beating eggs with sugar until fluffy not only incorporates air but also affects the texture of the cake, this could represent what one of the hidden layers is determining (there can be many hidden layers). These hidden layers enable the added complexity to be considered as part of the final prediction (in our case a cake).

The result, hopefully, is a perfectly baked cake—like the output of a deep neural network. However, instead of a delicious dessert, what a deep neural network produces are a prediction based on the 'recipe' it has learned from the data it was trained on.

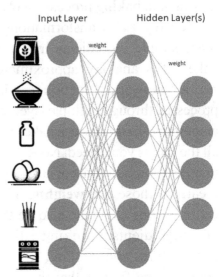

Figure 6 - Deep Learning
(Hidden Layer)

Deep learning involves adjusting these parameters or 'weights' across various layers, transforming the initial inputs into a final output. The term "deep" refers to the multiple layers that enable a computer to handle and interpret complex data with intricate subtleties.

But what if the cake doesn't taste good? In neural network terms, what if the prediction was inaccurate? You'd tweak the recipe—perhaps changing the amounts of sugar, flour, or baking time—and try again. This iterative process is known as **backpropagation**. It's basically a fancy way of saying that you did not get the result you

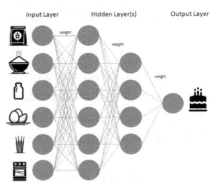

Figure 7 - Deep Learning
(Output Layer)

expected. You need the deep neural network to re-adjust the model weights (a little more flour and vanilla, a little less sugar, higher baking temperature, more whipping of the batter) and try again.

David Lloyd

This analogy helps explain why significant computing power is necessary. Each ingredient (input) can be adjusted endlessly in combination with others to achieve the optimal outcome. There are techniques that can be applied such as **Gradient Descent**, which methodically adjusts weights to minimize errors and refine predictions. In the case of our cake, decreasing perhaps the difference between the predicted and actual outputs (you used 1 cup of flour but ¾ of a cup was closest). Fundamentally, the goal is to determine how far off the model is from the ideals to find the optimal mix based on changes, getting smarter about the parameters (which includes the weights) of the ingredients (inputs) it sprinkles into decision making each time it runs. Effectively it's adjusting the parameters to optimize making a tasty cake.

As you read on you will see that there are several examples of more complex neural networks that use deep learning depending on the nature of the problem to be solved.

Now, let's explore other applications of neural networks:

Image Recognition: Your smartphone can unlock itself by recognizing your face, thanks to a neural network trained on millions of facial images. This ability enables devices to identify distinct facial features accurately. This approach often uses a Convolutional Neural network (CNN), covered shortly.

Handwriting Recognition: Delivery services use neural networks to interpret the scribbles on packages, ensuring they reach their correct destinations. This technology is also utilized in digital note-taking apps, improving their ability to convert your handwriting into typed text. This approach typically uses a Convolutional Neural network (CNN), covered shortly.

Language Translation: Companies like Google use neural networks to translate text between languages seamlessly, allowing for nearly fluent communication across linguistic boundaries. This approach typically users a Recurrent Neural Network (RNN).

Image Generation: Some neural networks can analyze art styles and generate new artworks that mimic famous artists, offering creative outputs based on learned styles and techniques. Images are created using a General Adversarial Network (GAN).

Recommendation Systems: Platforms like Spotify and Apple Music analyze your listening habits to recommend new music, using a blend of machine learning techniques to predict what you might enjoy based on your tastes and the preferences of similar listeners. These systems can also use a Recurrent Neural Network (RNN).

Image Identification: Identifying images or making predictions that don't use any "memory" of what happened in the past such as our home price example uses a Feed-Forward Neural Network (FNN).

As we dive deeper, we'll explore how these neural networks, through their complex and layered learning processes, tackle various challenges, from driving autonomous cars to managing intricate data across different platforms. These examples illustrate the broad and impactful applications of deep learning in our daily lives and beyond.

Feedforward Neural Networks (FNN), You Can't Go Back

Feedforward Neural Networks (**FNNs**) operate just as their name suggests; they process inputs through various layers to arrive at a prediction, without retaining any memory of past data. These networks are adept at tasks such as classifying whether an email is spam or analyzing sentiments in product reviews. They also excel in straightforward regression problems like estimating home prices based on specific features or diagnosing medical conditions from clinical data.

> *Regression looks to analyze relationships between input and output. For example, how one variable (price of a home) is dependent on another variable (sq. footage).*

Let's break down how a feedforward network could be implemented, using a real estate company as an example. Imagine you want to offer clients a predictive tool for home prices in a large city, which features over 5,000 properties for sale. Here's how you might go about it:

1. Identify Key Features: Determine which property attributes—such as type, square footage, location (postal code), age, and number of bedrooms and bathrooms—are most valuable to your clients.

2. Data Collection: Gather comprehensive data on homes sold in the area over several years, ensuring it includes all the relevant features and the last price sold for.

3. Data Preparation: Make sure the data is complete and consistently formatted across all features to prevent any input errors in your model.

4. Network Setup: Configure a neural network to process some of this data to predict home values based on the features you've identified.

5. Model Training: Use about 70% of your data to train the neural network, reserving the remaining 30% for model tuning and validation.

6. Validation and Testing: Use the reserved data to fine-tune and test your model, ensuring it generalizes well and makes accurate predictions.

7. Deployment: Integrate the model into your website, allowing prospective clients to input features important to them based on those you've defined and receive price predictions.

It's crucial to understand that feedforward networks treat each input independently—they do not track or analyze changes over time. If your data involves time series or historical patterns, a Recurrent Neural Network (RNN) would be more appropriate, as it's designed to handle such dependencies.

This example not only illustrates the practical application of feedforward neural networks but also underscores their utility in static input-output tasks where historical data isn't a factor. By leveraging these networks, businesses can provide valuable predictive insights that enhance decision-making.

Recurrent Neural Networks (RNN), I Remember Something

Recurrent Neural Networks (RNNs) shine when dealing with data where the present depends heavily on the past, making them ideal for tasks like language processing, speech recognition, and analyzing time-series or historical data.

RNNs possess a unique "memory" capability, allowing them to remember and use previous information in current decision-making processes. This memory is crucial when translating languages or generating text, as understanding the context from earlier in the conversation or text significantly enhances the accuracy and relevance of the output. Imagine reading this book: as you process each sentence, your brain recalls what you've read before, helping you understand each new piece of information in context. That's how an RNN operates—it keeps track of the narrative.

Consider the previous steps we've outlined for a FNN. The process for an RNN is similar but with a key difference: it feeds on historical data while using current data. This ability allows the network to "loop back" and consider previous inputs, enhancing its predictions over time.

The steps for setting up an RNN are parallel to those for an FNN but adapted to take advantage of the RNN's memory capabilities:

1. Identify Key Features: Just as with an FNN, determine which data points are most crucial. For stock price predictions, relevant features would include historical prices, trading volumes, and market conditions, interest rates, and so on.

2. Data Collection: Gather extensive historical data, ensuring it is comprehensive and well-documented to avoid gaps that could mislead the training process.

3. Data Preparation: Clean and preprocess the data to maintain consistency, which is crucial for training an RNN due to its dependency on sequential input.

4. Network Setup: *Configure the RNN to handle sequences of data, preparing it to analyze inputs in the context of their historical values.*

5. Model Training: Use a significant portion of your data to train the RNN, allowing it to learn from both the current and previous data points.

6. Validation and Testing: Continuously test the model using new data while comparing predictions against actual outcomes to ensure the model remains accurate over time.

7. Deployment: Implement the model in a real-world setting, such as a stock prediction tool on a financial website, enabling users to make informed decisions based on the model's forecasts.

Unlike FNNs, which treat each input independently, RNNs integrate past learning, making them particularly effective for predictions where past context enhances understanding or accuracy— like predicting the future price of a stock based on its historical performance. The RNN uses its learned patterns to forecast future behavior, continuously refining its predictions as new data becomes available.

By embracing the dynamic capabilities of RNNs, we can harness their power to predict, understand, and interact with the world in ways that mimic human memory and temporal awareness. This makes RNNs invaluable for applications requiring nuanced understanding over time, from financial forecasting to next-word prediction in texting apps.

Convolutional Neural Networks (CNN), I See Something

Convolutional Neural Networks (CNNs) are specially crafted to handle image data by basically overlaying a grid on images, akin to dissecting a puzzle piece by piece. Widely used in image detection for driverless cars, facial recognition, and medical imaging, CNNs leverage the structure of input-layers-output but introduce a concept called convolutional layers to master patterns crucial for image recognition. For example, a convolutional layer may be looking for certain aspects of an image, like the curve of a letter.

A CNN treats patterns in images numerically, easing the processing of simpler images, like black and white letters, and it grows more complex with color images. With increased image complexity, additional layers become necessary to capture finer details.

Let's explore how a CNN processes an image, using the letter 'Q' as an example, covered by a 3x3 grid. Each grid cell, or pixel, isn't analyzed in isolation; rather, groups of pixels are examined together.

A pixel represents the smallest unit of an image within a resolution. If you have a 600 x 800 pixel image it is composed of 480,000 pixels. Each pixel could represent over 16 million colors depending on the nature of the image. This is one reason why additional layers can be necessary in a CNN.

First, we take the input of an image. Perhaps we are looking at mail (the old-fashioned kind) and working to detect the postal code so we can route that bill properly to your home. Let's explore how a CNN processes an image, using the letter 'Q' as an example, covered by a 3x3 grid. Each grid cell, or pixel, isn't analyzed in isolation; rather, groups of pixels are examined together.

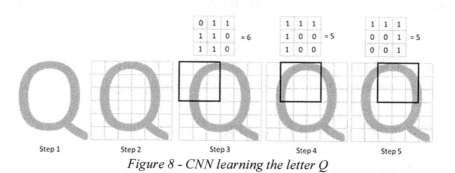

Figure 8 - CNN learning the letter Q

Next the first layer of the CNN consists of multiple convolutional filters or kernels. Each filter is small in size (in our example 3x3) as shown in the dark box grouped around the starting point of the image.

As the process continues a value is assigned to each block of the image that is being analyzed. This convolutional process works to determine patters from the image for things like edges, textures, or shapes. There are other more complex processes that occur but in general terms steps 3, 4, 5, and so on continue until the image has been reviewed and broken down into its smaller features and resulting computations.

Other actions such as pooling and flattening can be used with the goal of keeping the most meaningful data about the image. Using this data the CNN uses its neural network to classify the image into something more identifiable in this case, a letter of the alphabet, which was at some point classified by a human-in-the-loop. While this is a very simplified view of a complex process it's important to note that the size of the image, complexity, and even color depth will dramatically impact the speed in which an image can be trained and/or determined.

Using our previous example for Retinopathy, it would involve training a CNN with potentially thousands or hundreds of thousands of examples of both a healthy and unhealthy retina and a group of Ophthalmologists would classify the images of the retinas. The result would be a model that recognizes retina patterns in an image of a new patient predicting if they had Retinopathy.

Wrapping Up Deep Learning

Diving deeper into artificial intelligence, one encounters complex elements such as hidden layers, model design, and issues like overfitting and underfitting—each critical for refining AI's predictive capabilities. The design of neural networks, including decisions about the number of neurons and layers, largely hinges on a mix of problem complexity assessment and experimental fine-tuning. While automated methods for these decisions are emerging, they are far from standardized.

This part focused on providing some of the foundational aspects of machine learning that have propelled a multitude of applications, enriching our lives in remarkable ways. When querying OpenAI's ChatGPT about the essential AI terms everyone should know, it was gratifying to see that most have been covered in our discussions here, except for a few like bias, fairness, and ethics, which we will explore in relation to Large Language Models (LLMs) in Part 3. Each of these topics could itself fill books, underscoring their depth and significance.

Why is understanding this so crucial? As we've seen, deep neural networks utilize hidden layers which, while effective, hide the decision-making process. This lack of transparency means we often don't know how a prediction is exactly made, just that it is usually accurate—though not always. These systems aren't perfect; they inherit and can amplify the biases present in their training data, potentially leading to unfair outcomes influenced by culture, ethnicity, gender, and other factors. Such biases are particularly pronounced in large language models due to their extensive reach (the data they train on) and complexity. As we proceed, we'll discuss both the power and the pitfalls of LLMs, and consider strategies to mitigate these challenges, ensuring AI is used responsibly and ethically.

Now that we understand some of the basics, let's take a brief look at the path that brought us to where we are today.

Part 2: A Brief History of AI, The 1950s

"The good news is, we've discovered the secret of immortality. The bad news is, it's not for us."

--- Dr. Geoffrey Hinton, "Godfather of Deep Learning"

The 1950s were a crucible of change, marked by the return of global conflict with the Korean War, the coronation of Queen Elizabeth II, the persistent struggle for civil rights epitomized by the Montgomery Bus Boycott, and the cultural revolution ushered in by Elvis Presley. Amid these historic shifts, groundbreaking ideas about human intellect were quietly taking shape.

In the shadow of these larger events, from the fall of 1947 to the spring of 1948, Alan Turing was penning what would become a seminal work in the field of artificial intelligence. His paper, "Intelligent Machines," although initially dismissed by his contemporaries as a 'schoolboy's essay' and deemed unworthy of publication, laid down concepts that would endure to shape the future of AI. Despite the skepticism, Turing had the last laugh. His ideas not only founded the bedrock that would coin the term AI, but also secured his legacy as one of its pioneers.

"Schoolboy's essay," indeed.

During 1948, Turing submitted a report detailing the research he conducted after taking a year's sabbatical. Before this period, he had been engaged with the Automatic Computing Engine at the National

Physical Library, a project that proved frustrating due to its slow progress. This setback prompted Turing to take a leave of absence, during which he shifted his focus to a more theoretical domain. His supervisor described his new project as a machine "planned for work equivalent to that of the lower parts of the brain." Turing's explorations during this time were not just about what the machine could do but also whether it could "learn from experience."[4]

Intelligent Machines

In "Intelligent Machines," Alan Turing took a logic-based approach to exploring artificial intelligence, proposing that "intellectual activity consists of various kinds of search." This idea, though initially overlooked, would soon gain recognition among scholars. Turing's exploration didn't stop there; he also introduced the concept of a genetic algorithm inspired by the principle of 'survival of the fittest.'

Turing described the developmental stages of human intelligence as being analogous to the evolution of AI. He outlined three key components:

(a) The initial state of the mind, akin to a child at birth,

(b) The education received,

(c) Other experiences not classified as formal education.

Turing likened the human brain in its early stages to a blank notebook—ready to be written upon. Through 'evolutionary mutations' such as education, these 'child machines' gradually evolve into sophisticated adults.

This foundational view runs through the veins of almost all AI research today. Turing suggested that machines could possess

---

[4] Turing, Alan. "Intelligent machinery (1948)." *B. Jack Copeland* (2004): 400.

David Lloyd

neuron-like elements, similar to the human nervous system. Just as unorganized networks of neurons in infants eventually develop the capability to recognize parents' faces or communicate through basic expressions like crying or laughing, AI systems too can be 'trained' through interactions that reinforce or inhibit certain behaviors. These interactions effectively 'change' the AI into an optimized version of its initial form.

Turing masterfully translated these human developmental processes into the language of machines. He used a simple analogy: if a teacher commands a student to "do your homework now," the instruction holds only if the student accepts that "everything the teacher says is true." This condition applies equally to a human child or an AI 'child'. If the condition is met, the task is completed; if not, it remains undone, no homework.

Alan Turing's approach to artificial intelligence was profoundly logical, forming the backbone of what we now recognize as machine learning. He posited that just as children learn from the world around them, computers could be trained from their initial 'infant' state of disarray.[5] Turing's groundbreaking idea wasn't just that machines could be taught to search for answers or adapt to changes; he also envisioned a basic model for how such training could be systematically applied.

Turing later articulated a set of criteria to test a machine's ability to think, known as the Imitation Game.[6] This test involved three participants: a man, a woman, and an interrogator tasked with identifying them based solely on their responses to questions like, "How long is your hair?" The twist came when one of the participants was replaced by a machine. If the machine could

---

[5] Teuscher, Christof. "Foreword: Special issue on Alan Turing." *Evolutionary Intelligence* 5, no. 1 (2012): 1-2.

[6] Turing, Alan M. *Computing machinery and intelligence*. Springer Netherlands, 2009.

respond in a manner indistinguishable from a human, it was deemed to have passed the test.

This test, as Turing argued, delineated the intellectual capabilities of humans from the operational capacities of machines. The essence of the Turing Test evolved into a scenario where a human evaluator would converse with both a person and a machine designed to mimic human responses. If the evaluator couldn't reliably tell them apart, the machine was considered to have successfully passed the test.

Turing provocatively noted, "No engineer or chemist claims to be able to produce a material which is indistinguishable from the human skin," highlighting the futility of making a machine appear outwardly human when the real challenge lay in intellectual mimicry.

While the Turing Test remained a challenge yet to be conquered, it established a foundational framework for the pursuit of artificial intelligence. Turing's visionary concepts not only sparked immediate intrigue but also laid the groundwork for the complex AI developments that would follow. His work insists that the line between human and machine intellect is not just about thinking—it's about the subtlety and depth of thinking that truly matters.

The Mind

In the earliest days of computing, Alan Turing's groundbreaking theories on artificial intelligence were largely theoretical hamstrung by the limited computing power of the time. This bottleneck in technology is a pivotal reason why AI didn't hit its stride until decades later. Yet, Turing's initial explorations during the 1950s laid the groundwork for a future where his ideas would manifest into practical applications. As the world transitioned from wartime to a civilian focus, technology began to evolve in exciting new ways, spurred on by funding from entities like the Department of Defense at MIT's Lincoln Laboratory. It was here that some of the earliest practical attempts at AI took shape.

At MIT, Wesley Clark and Belmont Farley delved into the basics of 'neural networks'—self-organizing sequences designed for simple computers to mimic the brain's pattern recognition capabilities.[7] Their aim was to create a network that could perform tasks intelligently. We explored how neural networks became the bedrock of AI in the first part of this book.

Their initial task for these networks? Pattern recognition. These early computers were trained to recognize patterns by altering inputs and outputs—transforming the data in ways that modified the results. For example, a computer might initially mistake various features of the letter "Q" for an "O," but as it processed more characteristics, it would learn to correctly identify the letter "Q." This rudimentary mimicry of human brain processes demonstrated that computers, much like humans, could indeed learn.

Further contributions came from Lincoln Laboratory's Gerald P. Dinneen and Oliver Selfridge. Dinneen focused on teaching computers to clean up images—removing visual noise or enhancing clarity by analyzing pixel intensity (the individual dots of the making up the image) and contrasts. Meanwhile, Selfridge worked on enabling computers to recognize and categorize shapes from image features, like distinguishing a rounded corner from a pointed line.

In this context, Dinneen, Selfridge, Clark, and Farley were akin to 'preschool teachers' for early AI—using the neurological framework of the human brain as a template, they sought to construct its artificial counterpart, teaching it to discern letters and shapes as a human child might learn.

---

[7] https://dl.acm.org/doi/abs/10.1145/1455292.1455309 *W. A. Clark and B. G. Farley. 1955. Generalization of pattern recognition in a self-organizing system.*

These scholars at Lincoln Laboratory were not just building tools; they were nurturing a nascent intelligence that could see, think, and eventually speak. Their work laid a foundation that many others would build upon, though not without debate. As research progressed, a divergence in approaches emerged, signaling the beginning of a vibrant, ongoing debate on the best paths forward in AI development—a debate that continues to shape the field today.

The Maze

By 1955, the concept of Alan Turing's 'thinking machine' had been transformed into reality—not through the replication of neural networks, but through the pioneering work of researchers Allen Newell and Herbert Simon from Carnegie Mellon and RAND. Drawing directly from Turing's theory that intellectual activity is essentially various forms of searching, Newell and Simon crafted a machine that demonstrated how simple units of logic could interact to exhibit behavior similar to human reasoning.

In stark contrast to their MIT contemporaries, who focused on simulating neural networks, Newell and Simon adopted a different metaphor for human decision-making: a maze. They theorized that human problem-solving operates through logical sequences—seeing a symbol, interpreting it, and making decisions based on that interpretation. Directions like 'turn left,' 'stop,' or 'slow down' are all part of the maze of choices that lead to a solution. Their perspective suggested that creative problem-solving was not merely the product of complex neural networking but rather navigating through a labyrinth of symbolic decisions.[8]

Their creation, the Logic Theorist, epitomized this theory. It could prove mathematical theorems with the same proficiency as a seasoned mathematician. Given specific instructions, the Logic Theorist followed routines and sequences to discover the correct

---

[8] http://shelf1.library.cmu.edu/IMLS/MindModels/creativethinking.pdf

David Lloyd

approach for solving proofs—essentially showing how conclusions follow logically from premises.

> *Example of a proof: adding two even numbers always result in an even number, or non-mathematically, all oceans have salt water.*

The Logic Theorist would solve it. Perfectly.

This machine was a monumental step in the evolution of AI. The Logic Theorist was the first true application of a machine performing complex tasks in a manner analogous to human thought processes, marking a significant milestone in the field of artificial intelligence. Its success was not just a demonstration of machine capability but a profound insight into the potential for machines to think logically, to solve problems, and to navigate the mazes of intellectual challenges just as humans do.[9] Not an average human, either: Logic Theorist could do math like a professional.

It could be *trained.*

No longer just one of Turing's theoretical "child-machines"—which, as he pointed out in "Intelligent Machines," didn't need to have legs or even eyes—the Logic Theorist had evolved, becoming what might be described as a "teen-machine."

Meanwhile, Herbert Simon and Allen Newell, heralded as the fathers of America's first Thinking Machine, weren't resting on their laurels. Along with Cliff Shaw, a key collaborator on the Logic Theorist, the trio was already pushing boundaries. They went on to develop the General Problem Solver, an ambitious extension of their original tool. This new machine represented another significant

---

[9] Gugerty, Leo. (2006). Newell and Simon's Logic Theorist: Historical Background and Impact on Cognitive Modeling. Proceedings of the Human Factors and Ergonomics Society Annual Meeting. 50. 880 884.

advance in the capability of machines, designed to tackle a broader range of formal problems and proofs.

Yet, despite these advancements, the development of what would later be widely recognized as artificial intelligence was still flying under the public radar. The General Problem Solver, while a leap forward, was confined to solving structured problems that could be explicitly defined or formalized, limiting its immediate impact outside academic circles. The journey of AI from a secluded lab experiment to a central player in technology's narrative was just beginning—a slow burn that would soon catch fire and capture the global imagination.

Dartmouth

The 1956 Dartmouth Conference—officially dubbed the Dartmouth Summer Research Project on Artificial Intelligence—marked a pivotal juncture where the physical proof of humanity's venture into the intelligent unknown met the brightest minds capable of steering its course.

What was initially planned as an intimate gathering transformed significantly as the word spread. Anticipated to host just 11 attendees, the event instead attracted a crowd of 47 curious minds. Although modest compared to mainstream media spectacles, this turnout was notable for what was essentially an extended brainstorming session on nascent technology. This small but engaged group had perhaps sensed they were on the cusp of history, participating in what would later be celebrated as the birthplace of artificial intelligence.

In reality, artificial intelligence was not being born but rather evolving rapidly. Turing's metaphorical 'infant data' was maturing. Innovators like Clark and Farley had devised machines capable of thinking, while Dinneen and Selfridge engineered machines that could 'see.' Newell and Simon had already demonstrated that machines could tackle complex tasks through their creations, the Logic Theorist and the General Problem Solver. The field was growing, pulling along a dedicated cadre of scholars intrigued by its potential.

John McCarthy, a math professor at Dartmouth and the orchestrator of this historic gathering, used a Rockefeller Grant to bring together these minds in Hanover during what would be the wettest summer the area had seen in years. In their original proposal, McCarthy and his colleagues ambitiously outlined their goal to explore how machines might use language, form abstract concepts, solve problems typically reserved for humans, and even enhance their own capabilities'[10]

The conference did more than foster discussion—it crystallized the field's identity. While some had referred to this burgeoning area as cybernetics and others as computational intelligence, it was at this gathering in New Hampshire that a consensus was reached. The field would be known henceforth as **artificial intelligence**, a term coined by McCarthy, capturing the essence of what Turing had once described as artificial life.

Notable attendees included Herbert Simon and Allan Newell, key figures from RAND and Carnegie Mellon; Oliver Selfridge from Lincoln Lab; and Marvin Minsky, who would later establish MIT's Institute for AI. Mathematician John Nash was also present, known for his 'beautiful mind' and his foundational work in game theory, which resonated well with Newell and Simon's approach to symbolic problem solving.

The Dartmouth Conference proved that AI was not just a fleeting academic fancy, but a robust field poised to stride into practical applications. With the combined intellectual power of the nation's

---

[10] McCarthy, John, Marvin L. Minsky, Nathaniel Rochester, and Claude E. Shannon. "A Proposal for the Dartmouth Summer Research Project on Artificial Intelligence, August 31, 1955." *AI magazine* 27, no. 4 (2006): 12-12.

top scholars in one room, that rainy summer in Hanover was a watershed moment. It propelled the field of artificial intelligence from a theoretical pursuit into a tangible, dynamic force, setting the stage for a future where machines not only mimicked human thought processes but began to enhance them. This seminal gathering tipped the scales, shifting the trajectory of technology from speculative fiction towards tangible reality—akin to moving from the realm of frog eggs to that of DNA-engineered dinosaurs.

Teaching Computers To Read

Part of recognizing language is recognizing letters. Optical character recognition (OCR) was created to teach machines this skill.[11] Like in the work of Dinneen and Selfridge, machines could be taught to read by interpreting shapes, then numbers. The technology developed fast, as number recognition is useful.

The initial focus was in developing ways that computers could understand patterns of typically written letters and numbers. The importance of that included reading postal codes from mail, remember mail? This aided the postal service in sorting mail based on postal code destination. Also, eventually the use was applied to reading numbers and other digits off the bottom of checks, remember those as well?

Earlier in the use of OCR systems the focus was on determining handwritten letters and matching these to a predefined set of examples to determine the likely character being represented. This was a challenge as handwriting styles varied significantly, but in the case of OCR for use in banks for check routing, the fact that numbers were standardized made the pattern matching accuracy far higher, enabling this to be used starting in the late 50's.

---

[11] Nilsson, Nils J. *The quest for artificial intelligence*. Cambridge University Press, 2009.

Artificial intelligence was on its way, and we were just along for the ride.

The 1960s: Ready for Liftoff

"Once delivered from the limitations of biology, we will be able to decide the length of our lives--with the option of immortality-- and choose among other, unimagined capabilities as well."

--- Marvin L. Minsky

If there was ever a decade that showcased human triumph, it was the 1960s. Humanity conquered the moon just as we mastered our own skies. In an era where time was synonymous with money, one didn't need to be Neil Armstrong to cross the Atlantic on a supersonic jet. To put it in perspective, the calculator in your desk drawer (yes, before it was replaced by smartphones) boasted more computing power than the technology aboard Apollo 11.

The 1960s were marked by iconic moments: Martin Luther King Jr.'s "I have a dream" speech, the first human heart transplant, Neil Armstrong's lunar steps, and the birth of the Pop Tart. Musically, we transitioned from Elvis and doo-wop to the Beatles, Jimi Hendrix, and The Supremes.[12]

Things were shaking.

Yet, the '60s for AI were as dynamic as they were for the culture at large. The eagle had landed, and artificial intelligence was gearing up for its own monumental ascent.

The brilliant minds that convened at the Dartmouth Conference in the previous decade were now steering AI through its formative steps

---

[12]

toward becoming the powerhouse it is today. From the theoretical seeds planted by Turing's intelligent machines sprouted real systems capable of complex operations, guided by America's brightest problem solvers. Universities like MIT, Stanford, and Carnegie Mellon became breeding grounds for innovation—fostering growth that was as rapid and transformative as fungi in the wild, cementing the foundational roots of a profound technological evolution.

In this vibrant ecosystem, artificial intelligence found its rhythm, setting the stage for a revolution that would reshape the world.

DARPA's Big Save

At the heart of this seismic shift in AI was a mechanical revolution. In Turing's era, computers had a significant limitation: they couldn't save information. The pioneering machines that ran the Logic Theorist and the General Problem Solver could execute commands but couldn't retain them. Using Turing's metaphor, it was as if these computers were like infants learning a new word only to forget it instantly.

This inability to store information was a major barrier that delayed the realization of Turing's visionary ideas. Additionally, the exorbitant cost of computers—which could run into hundreds of thousands of dollars monthly—posed another formidable challenge. Icons like Bill Gates borrowed his, while Newell and Simon capitalized on RAND's to fund theirs.[13] The landscape began to change dramatically in 1963 with the entrance of the Defense Advanced Research Projects Agency (DARPA).

DARPA's involvement was pivotal for several reasons. Its financial support did more than just sustain the field of AI; it empowered it. The agency provided funds with no strings attached, allowing computer scientists at MIT, Stanford, and Carnegie Mellon the freedom to

---

[13] https://www.cmu.edu/simon/what-is-simon/history.html

innovate in "excellence in information processing in whatever fashion we thought best," as Allan Newell put it. This unfettered support proved to be a crucial catalyst, nurturing a burgeoning field that was ready to leap from theoretical frameworks to real innovations.[14]

This era of generous, unrestricted funding helped AI researchers to overcome the limitations of earlier technologies, setting the stage for the development of computers that could remember as well as think—a fundamental step towards the AI systems we interact with today. The 1960s, therefore, were not just a time of cultural upheaval but also a period of profound technological empowerment that would define the future of artificial intelligence.

In the 1950s, the groundwork was laid for machines to see and optimize outcomes. The 1960s took these capabilities further, extending them into the realm of thinking, albeit in a rudimentary form. For machines to truly think, they needed the ability to learn, and the advancements in learning during this decade were profound, spanning activities from playing checkers to engaging in conversation.

The first notable milestone came in 1962, when Arthur Samuel's IBM 7094 computer faced off against checkers master Robert Nealey in a match broadcast nationwide. This event wasn't just a public spectacle; it symbolized a significant leap in machine learning, demonstrating that machines could not only execute pre-programmed instructions but also adapt and improve through experience.

As the post-war era ushered in steady funding, the 1960s became a golden age for AI development. Scholars and researchers, flush with resources and driven by curiosity, pushed the boundaries of what machines could do. Whether it was navigating the complexities of a

---

[14] Fikes, R., & Garvey, T. (2020). Knowledge Representation and Reasoning — A History of DARPA Leadership. *AI Magazine, 41*(2), 9-21. https://doi.org/10.1609/aimag.v41i2.5295

checkers board or simulating basic conversation[15], AI was beginning to show glimpses of a future where machines could perform tasks previously reserved for the human mind.

This decade of innovation, fueled by enthusiasm and significant financial backing, set the stage for AI to evolve from a series of intriguing experiments into a field of study that promised to revolutionize how we interact with technology from theory to a tangible agent of change in everyday life.

LISP's List

The initial push to bring AI into the public eye focused significantly on understanding human language and enhancing self-learning capabilities. The 1960s, a decade rich in technological innovation, particularly excelled in advancing both neural networks and symbolic processes, with language processing taking center stage.

> *While covered in part two, think of a neural network like the human brain with neurons connected by many synapses and symbolic processes developed through programming languages (LISP for example).*

The 1960s didn't just broaden the applications of AI; they also provided the essential tools for its creation. John McCarthy, who was pivotal in conceiving List Processing (LISP) alongside the Logic Theorist in 1958, was deeply influenced by his interactions with luminaries like Simon and Newell.

Developed in the wake of the Dartmouth Conference and alongside the formation of the AI group at MIT, LISP became one of the oldest programming languages still in use today, a testament to its foundational role in AI development.

---

[15] https://www.ibm.com/ibm/history/ibm100/us/en/icons/ibm700series/impacts/

David Lloyd

John McCarthy, when asked about what was needed to kickstart their project with Marvin Minsky, famously quipped they required "a room, two programmers, a secretary, and a keypunch." They were granted all these resources, along with the support of six graduate students in mathematics.

In practice, LISP significantly simplified the coding process compared to the original Turing machines. It adeptly managed complex AI tasks like knowledge representation and interactive development, allowing programmers to execute code and immediately see results— a stark contrast to earlier methods where one had to wait and check if the code had succeeded or failed.

LISP's efficiency lay in its simplicity. For instance, the command (+ 3 7) in LISP is straightforward: it interprets a function (+) and its arguments (3 and 7) to return a value of 10.

While this is a simplified overview, the principle holds: LISP applied logical constructs to teach old machines new tricks and, thanks to advancements in computing power, enabled them to remember these lessons. Although it had its quirks (and indeed induced a sort of parenthesis phobia among early programmers), LISP endured long past its inception because, quite simply, it worked. This practical effectiveness is why, despite its initial peculiarities, LISP remains a cornerstone in the history of AI programming.[16]

STUDENT, a computer program crafted by Daniel Bobrow, serves as a fascinating illustration of the early applications of Natural Language Processing (NLP), a subset of Artificial Intelligence aimed at enabling computers to comprehend, interpret, and generate human language. It was designed specifically to tackle the kind of word problems you might encounter in a high school algebra class, demonstrating how a computer could deconstruct verbal expressions into their simplest forms to solve complex problems.

---

[16] http://www-formal.stanford.edu/jmc/history/lisp/node6.html

Let's consider a hypothetical problem from 1964, the year the world was introduced to the Pop Tart:

Susan has twice as many Pop Tarts as Kevin. Combined, they have a total of 18 Pop Tarts. How many Pop Tarts do they each have?

STUDENT would then read the text, which might use S to indicate the number of Pop Tarts Susan has, and K for the number Kevin had. It would then determine, based on the text that Susan had twice as many pop tarts $(S \times 2) + K = 18$. From there it would carry out the formulas it had built.

In essence, STUDENT wasn't just a tool for pattern recognition—it approached pattern recognition in a distinctly 'human' way, making it an early bridge between straightforward computation and the nuanced understanding of language.

This progression in AI was significant. While STUDENT was focused on solving algebraic word problems, it set the stage for more advanced applications like ELIZA, which would soon venture beyond mathematics into the realm of conversational AI. ELIZA, often considered the precursor to modern chatbots, took the capabilities demonstrated by STUDENT and expanded them into the domain of interactive communication, acting as a virtual therapist and marking a significant milestone in AI's journey into human-like interactions.

Thus, while STUDENT and ELIZA were early forays into the world of AI, they heralded a new era of technological interaction—paving the way for the sophisticated chatbots and virtual assistants that are now a staple of modern digital experiences.

David Lloyd

ELIZA, My Fair Lady?

Then came ELIZA.

Unlike the earlier, more numerical ventures into NLP or Optical Character Recognition (OCR), ELIZA represented a different kind of breakthrough that wasn't immediately recognized for its practical utility. Created by Joseph Weizenbaum, a German-American professor with a dramatic personal history of fleeing Nazi Germany in 1936 and serving as a meteorologist during World War II, ELIZA emerged from a rich backdrop of resilience and renewal.

Weizenbaum, who joined MIT just two years after the cinematic release of "My Fair Lady" in 1964—a fact that whimsically coincides with his creation's name. ELIZA however, was actually named after the character Eliza Doolittle from "Pygmalion" which set the stage for future developments in NLP and chatbot technology. Much like her namesake, ELIZA could be "taught" and refined to improve her linguistic abilities over time. However, unlike Eliza Doolittle's transformation into a polished member of society, ELIZA's role was more akin to that of a psychologist.

ELIZA operated through a script known as DOCTOR, which mimicked the conversational style of a psychotherapist. Yet, the technological constraints of the time rendered her dialogue style somewhat stark and blunt. Due to the peculiarities of the MIT computer system, where question marks were used as line delete characters, ELIZA's responses lacked the softness typically associated with therapeutic discourse. This led to a disconcerting, almost eerie manner of interaction, where her bluntness contrasted sharply from what one might expect with a nurturing human therapist.

An illustrative example from Weizenbaum's 1966 paper showcases ELIZA's responses (all in capitals) to human inputs, highlighting both the groundbreaking nature of her design and the starkness of her digital interactions.[17] This closeness made ELIZA not just a technical marvel but also a subject of fascination and debate within the AI community, as she blurred the lines between machine processing and human-like conversation.

">You are like my father in some ways.
WHAT RESEMBLANCE DO YOU SEE

>You are not very aggressive but I think you don't
want me to notice that.
WHAT MAKES YOU THINK I AM NOT VERY
AGGRESSIVE

>You don't argue with me.
WHY DO YOU THINK I DON'T ARGUE
WITH YOU

>You are afraid of me.
DOES IT PLEASE YOU TO THINK I AM
AFRAID OF YOU"

ELIZA was undeniably a breakthrough in artificial intelligence, but whether she was a comforting presence is another story entirely. She certainly didn't soothe the nerves of her creator, Joseph Weizenbaum, who, despite ELIZA's success, became one of AI's most vocal critics.[18] Weizenbaum was particularly troubled by the ease with which people attributed human-like emotions to ELIZA during interactions, a phenomenon that highlighted the deep psychological impact such technology could wield.

ELIZA, like STUDENT, operated by applying pattern rules to deconstruct sentence inputs and generate responses. Weizenbaum

---

[17] https://dl.acm.org/doi/pdf/10.1145/365153.365168
[18] https://www.nytimes.com/1977/05/08/archives/experts-argue-whether-computers-could-reason-and-if-they-should.html

described ELIZA not just as a program or an interpreter but as an actor delivering lines scripted in advance.[19] This triple role encapsulated the complexity and the potential of ELIZA to mimic human conversation—albeit with a disconcerting bluntness due to the limitations of her programming environment at MIT, where punctuation marks could alter command executions.

Despite her blunt communication style, ELIZA marked the genesis of what we now recognize as chatbots. Whether the evolution of chatbots is beneficial remains a topic of debate and contemplation. As we continue to interact with these digital entities, the line between tool and companion blurs, echoing Weizenbaum's early reservations about the path AI might take. Thus, ELIZA stands not just as a technical achievement but as a harbinger of the complex ethical and psychological landscapes that we navigate in the ongoing development of artificial intelligence.

Good Neighbors

Interested students flocked to artificial intelligence from fields across campus, especially in areas like statistics and math. This was where Evelyn Fix and Joseph Hodges emerged with the concept for **K-Nearest Neighbor** (KNN), which would go on to become a key component of AI. Although Fix and Hodges introduced the idea in 1951, it wouldn't be tested on a real machine until Thomas Cover in 1967.[20] The delay was thanks to the pesky problem of computer memory, still very limited at the time.

K-Nearest Neighbor (KNN) is a concept that, at its core, calculates the likelihood of a piece of data belonging to a particular class or possessing a specific attribute. This algorithm has become a cornerstone of modern recommendation systems—you know, the kind that suddenly flood your browser with sunscreen ads after

---

[19] https://dl.acm.org/doi/pdf/10.1145/363534.363545
[20] https://isl.stanford.edu/~cover/papers/transIT/0021cove.pdf

you've just booked a flight to Mexico. Here's how it works: the algorithm notices your search for flights, categorizes you as a 'traveler,' and then tailors its recommendations accordingly. Beyond just influencing your shopping experiences, KNN is also employed in more critical areas like evaluating health risks from gene expression or assessing someone's creditworthiness for a loan.[21]

> *As we explored in part 1, KNN functions by classifying data based on known input and expected output. For example, if the input is booking a flight to sunny locations, the output might suggest destinations like Belize or related products geared towards warm climates.*

Despite its utility, KNN, much like LISP, is not without its limitations. However, the very fact that it remains in use today speaks volumes about its effectiveness. In the realm of AI, where newer technologies constantly push the boundaries of what's possible, the persistence of methods like KNN underscores an important lesson: newer isn't always better. Indeed, when it comes to developing intelligent systems, the simplest and most transparent approaches often lead to the most reliable and understandable solutions.

> *This recognition that some of the oldest algorithms in AI are still relevant today challenges us to think differently about progress and innovation. It isn't always about the latest technology like generative AI; sometimes, it's about using established tools in new and imaginative ways.*

Building Blocks

Terry Winograd's creation of SHRDLU for his PhD dissertation at MIT marked a significant advancement in **Natural Language Understanding** (NLU). SHRDLU was essentially an AI preschool

---

[21] https://www.ibm.com/topics/knn

teacher but for machines, training them to interact with blocks through typed commands. Users could instruct SHRDLU to move, stack, or clear blocks, and then question it about the positions of these blocks. This interaction wasn't just child's play; it demonstrated sophisticated artificial intelligence capabilities, hinting at the real-world potential of robotics to perform similar tasks.

The background of SHRDLU traces back to the original Linotype keyboard, which was arranged not in the QWERTY layout we're familiar with today but by the frequency of letter usage, resulting in the sequence ETAOINSHRDLU. Linotype machines lacked a backspace key, so typographical errors were often masked by a random string of keystrokes—typically "SHRDLU" used to fill out the line. This accidental string of characters became emblematic of a kind of technological mishap.

Winograd recalls that he chose the name "SHRDLU" possibly from a science fiction tale he read in his youth, penned by Frederic Brown and titled after the familiar Linotype sequence. In Brown's story, a Linotype machine starts absorbing what is typed on it and attempts to dominate the world—a narrative not unlike many science fiction warnings about the perils of overreaching technology. The crisis in the story is averted only when every book on Buddhism ever written is fed into the machine, a metaphor for tempering technology's reach with human wisdom.[22]

This story, while a cautionary tale about technology, highlights a broader theme. Even as Linotype machines revolutionized typesetting by automating a laborious process, they still required human oversight to correct their 'SHRDLU' errors. Similarly, SHRDLU's abilities to manipulate and interpret block arrangements demonstrated AI's potential, but also underscored the necessity of human guidance and oversight. Through SHRDLU, Winograd not only showcased the possibilities of AI but also subtly reminded us of the persistent need

---

[22] https://hci.stanford.edu/~winograd/shrdlu/name.html

for human interaction in the digital age. It's a direct precursor to the AI assistants (like Siri) we use today.

The 1970's: Expert Solutions

"Whereas the short-term impact of AI depends on who controls it, the long-term impact depends on whether it can be controlled at all."

--- Stephen Hawking

The 1960s gave us many things, among them the first pangs of discontent with the intellectual use of machines. Could a robot learn to speak? Yes. Should it?

Well...

The 1970s marked a distinct turn towards practical applications in the field of artificial intelligence. Building on the foundation of natural language programming from the 1960s, this decade introduced the concept of "expert systems." These systems, like the General Problem Solver before them, were designed to perform complex tasks previously reserved for humans. For Allan Newell and Herbert Simon, it was solving mathematical problems, while for Daniel Bobrow, it involved tackling the type of math typically done by human experts. The '70s saw these expert systems branching into new domains, enhancing their practical utility significantly.

Prior to this, though, was Shakey.

Shakey The Robot

Shakey emerged from the intellectual environments fostered by John McCarthy and Marvin L. Minsky, who were instrumental in the development of LISP and the establishment of significant AI labs at MIT and Stanford. Shakey transcended the traditional confines of computer-bound AI, venturing into the physical world as a mobile robot capable of interpreting and interacting with its environment.

Shakey was truly the first of his kind—a sizable robotic entity equipped to follow instructions, learn from its surroundings, and plan routes. If earlier AI could see and think, Shakey could also move, albeit on wheels and with somewhat limited stability.

Shakey's functionality was divided into levels. The base level involved the mechanics and software that enabled basic movements like rolling and tilting. More advanced capabilities included the ability to open doors, navigate rooms, and move objects, all facilitated by visual inputs from cameras. At its highest operational levels, Shakey stored sequences necessary for completing tasks and had a 'nervous system'—a sophisticated program that determined which commands to execute and when.

What set Shakey apart was his autonomy; he did not require a human operator for decision-making. Whether turning on lights or navigating obstacles, Shakey's catlike sensors detected presence and adjusted actions accordingly.[23]

Shakey's impact extended beyond the laboratories and academic circles. He captured the media's attention, becoming a public symbol of what AI could achieve. For the first time, the broader American public was gaining insight into the advancements that had been quietly evolving behind closed doors. Shakey wasn't just a collection of codes; it was the physical embodiment of the era's "intelligent machines," showcasing a significant leap from theoretical constructs to tangible, interactive technology.

Benjamin Kuipers, one of Shakey's developers, remarked that the robot was ahead of his time, a sentiment that captures the pioneering spirit of the era's AI advancements. Shakey not only pushed the

---

[23] Center, Artificial Intellgence. "Shakey the robot." (1984)

boundaries of what machines could do but also expanded the public's imagination about the future of technology

"Even today," he says, "When robotics is being taught to high school students, and computing and sensors cost almost nothing, most robots in labs and companies do not have the AI capabilities that Shakey had in the 1970s."[24]

Shakey had a bit of a totter (hence the name), and he took his time, especially with baseboards. Still, his existence was enough to inspire a new generation of Americans: not the least of them Bill Gates.

DENDRAL, MYCIN, and Prospector's Prospects

Like the intricate mathematical proofs tackled by the Logic Theorist, complex chemical compounds were traditionally the realm of (human) experts—until DENDRAL. Developed at Stanford University by Edward Feigenbaum, Joshua Lederberg, and Bruce Buchanan, and buoyed by DARPA funding, DENDRAL revolutionized the approach to organic chemistry.

DENDRAL was to chemistry what the Logic Theorist was to mathematics: it transferred the ability of human logical problem-solving into the digital hands of a machine, teaching it to learn from each task it performed. DENDRAL exemplified the idea that "knowledge is power" in the most literal sense. Operating on an 'if/then' premise, it navigated through vast domains of data, pinpointing specifics relevant to its chemical analyses and using these findings to identify solutions, all while learning about chemistry—arguably with more zeal than many high school students.

---

[24] Kuipers, Benjamin, Edward A. Feigenbaum, Peter E. Hart, and Nils J. Nilsson. "Shakey: from conception to history." *Ai Magazine* 38, no. 1 (2017): 88-103.

David Lloyd

Like its predecessors, the Logic Theorist and the General Problem Solver, DENDRAL marked a significant leap forward by demonstrating that computers could not only solve complex problems but could also acquire and apply new knowledge. It went beyond solving equations; it delved deep into its subject matter, creating a separate knowledge base that it used to enrich its understanding and enhance its problem-solving capabilities.

Moreover, DENDRAL was not just a theoretical marvel—it had practical applications. It provided 'real world' solutions to real-world problems in chemistry. The primary users of DENDRAL were not its developers but actual chemists engaged in active research. These professionals were less concerned with the intricacies of the programming or the mechanics of the AI. What mattered to them was its effectiveness and reliability in performing complex chemical analyses.

In this way, DENDRAL not only pushed the boundaries of what artificial intelligence could achieve but also demonstrated how AI could be seamlessly integrated into professional fields, enhancing human capabilities rather than merely replicating them.[25]

At the same time as DENDRAL was making waves in organic chemistry, another significant piece of AI, MYCIN, emerged as what some called the 'granddaddy of expert systems'. Developed in the 1970s, MYCIN was designed to specialize in the field of antibiotics, specifically targeting infectious diseases.

MYCIN was a pioneering example of how AI could mimic the decision-making capabilities of a human expert. It could analyze an infection, trace it back to its source, and recommend a treatment plan. Equipped with over 600 rules in a yes/no framework, MYCIN

---

[25] https://deepblue.lib.umich.edu/bitstream/
handle/2027.42/30758/0000409.pdf%3Bjsessionid%
3D70A227C7B0C5A1B04090514F8977501C?sequence%3D1

methodically narrowed down its diagnostics to pinpoint the appropriate intervention—capabilities that, while not as quick as a human expert, were impressively accurate in identifying treatments for diseases like meningitis.

The implications of MYCIN extended beyond its immediate medical applications. It demonstrated that machines, when given adequate information and tools, could make decisions and function in ways previously considered the exclusive domain of human experts. This ability to replicate and even enhance expert-level decision-making at a fraction of the cost and time was revolutionary.

The era's socio-economic context underscored the need for such practical and efficient solutions. With the United States recently off the gold standard, winding down the Vietnam War, and grappling with economic issues like stagflation and rising consumer prices, efficiency and practicality had become paramount.

It was in this climate that the next expert system, PROSPECTOR, was developed. Tailored for the field of geology, PROSPECTOR utilized a version of LISP called INTERLISP to perform the 'expert job' of identifying potential sites for drilling and mining. It would analyze ore deposits and digitized contour maps to pinpoint locations likely to yield valuable resources.

Remarkably, PROSPECTOR performed these tasks with near-expert accuracy. It employed a sophisticated system of yes/no queries—where 'E' represented empirical findings and 'H' stood for hypotheses—mirroring a complex game of 'twenty questions'. Through this method, PROSPECTOR could guide users to the most promising drilling sites, often matching or even surpassing the recommendations of human geologists in accuracy.

This ability to integrate deep knowledge into practical decision-making tools showcased a pivotal shift in AI from theoretical exploration to real-world application, emphasizing AI's role in driving not just technological but also economic advancements. As such, these expert systems not only transformed their respective fields but also set a precedent for future AI applications to be both profit-driven and practical, fitting seamlessly into the broader narrative of efficiency and innovation that defined the era.[26]

PROSPECTOR used **Bayes'** model of fuzzy logic for the confidence it had in its classification of 'this rock suggests oil' or 'not at all close to oil'. Fuzzy logic was based on the premise that not everything can be a simple yes or no. It's unique in how it assists computer systems in situations and decision making where the answer is unclear. Unlike DENDRAL or MYCIN, PROSPECTOR represented aspects of the real world through associations (rock, mineral, fossil). It was also developed with a real person in mind.

PROSPECTOR was designed to look for oil, but in some ways it was the next stage of Optical Character Recognition (OCR). It used some of the same image recognition technology approaches applied to characters, but now used the geologists contour map and gave the user its confidence in its decision, not simply yes or no. Early OCR models had this as well: for example, being 90% sure a written character was a D, and 10% sure it was a 0.

Now, PROSPECTOR could look at a contour map and make meaning from the curves: meanings on which a human expert would agree. In this sense, it was a step farther than earlier attempts. PROSPECTOR didn't just categorize landscapes and contours, the way Selfridge's image recognition machines recognized shape. PROSPECTOR looked and inferred. It made a choice – and it did so all by itself.

---

[26] http://www.computing.surrey.ac.uk/ai/PROFILE/prospector.html#Inference

Frames

The 1970s not only revolutionized the practical application of machines but also introduced new paradigms for interpreting and representing knowledge. Among these innovative frameworks was Marvin Minsky's concept of "frames," which he detailed in his influential 1974 book, "A Framework for Representing Knowledge." This work built on his earlier contributions, like his 1969 seminal text on "Perceptrons," which laid the groundwork for the burgeoning field of neural networks and pattern recognition.

Minsky's focus in the 1970s shifted toward understanding the cognitive architecture of the mind, beginning with its foundational elements. He proposed that our understanding of the world is constructed around "frames" — mental structures that represent stereotypical situations. For instance, when you walk into a room, you have a preconceived framework that includes walls and a door.

According to Minsky, each frame is equipped with various types of information: instructions on using the frame, expectations of what might happen next, and strategies for what to do if these expectations aren't met. Within these frames, there are "terminals" — slots that are filled with default information. In the context of a room, these might include elements like windows, a rug, or a chair.

However, consider a scenario where some typical elements are missing: the room has concrete walls, a bare floor, no windows, and no chairs, but there are bars on the walls. This simple marker might lead you to conclude you're in a jail cell, particularly if you move to the next room and see tables, benches, and people in jumpsuits. This related but distinct scenario forms a new "frame" within your cognitive system.

Through this system of frames, terminals, and transformations, Minsky illustrated how complex, innate knowledge could be deconstructed into simpler, computable parts that a machine could understand. A machine doesn't need to comprehend the societal implications of imprisonment; it merely needs to recognize the visual cues that signify a jail.

Minsky's theories established the fundamental building blocks of AI: recognizing cues, adjusting when anomalies arise, and recalculating based on new information. For example, upon hearing "10-year-old Johnny is playing with his trucks," a human can infer potential locations like a park, a school, or home. Adding "Then, his mother pushed open the law firm door," forces the brain to reevaluate the scene, perhaps concluding that Johnny is at his parent's workplace.

These rapid cognitive adjustments are everyday feats for the human brain, yet they represented a profound challenge and opportunity for AI development during Minsky's era. His work harkened back to earlier concepts like "semantic nets," introduced by Cambridge scholar Richard H. Richens in 1956, demonstrating how contemporary AI was continually informed by its academic lineage. Minsky's frames not only advanced our understanding of how AI could mimic human thought processes but also underscored the intricate interplay between cognition, representation, and machine learning.[27]

A simple semantic net might be:

    [Cat] --(has)--> [Four legs]
    [Cat] --(is)--> [Animal]
    [Cat] --(can)--> [Meow]

In Minsky's world, the frame for 'cat' might have default terminals and markers like four legs, animal, can meow. A transformation: the cat can bite. It isn't in the logic, but under Minsky's knowledge 'frames', the machine can assume the creature is still a cat.

---

[27] https://www.mt-archive.net/50/MT-1956-Richens.pdf

It was rudimentary – Minsky only conceptualized frames, not built them – but the lens of artificial intelligence was becoming sharp and wide. The field was beginning to cut its teeth.

Speech

Recognition of speech, a cornerstone of contemporary artificial intelligence, was beginning to take shape, though full mastery would develop later. The groundwork during this period involved **Hidden-Markov Models** (HMM), which had been utilized in mathematics and engineering well before their application in artificial speech. An HMM is adept at predicting missing inputs based on known outcomes—like deducing which of two coins, one biased towards heads, was tossed based on the toss result.[28]

With speech, the application of HMMs diverges slightly. Consider how you hear the syllables in a word, such as "cuh-ah-tuh" for "cat." These individual sounds are measurable, and though the word itself might initially be unknown, its component sounds are detectable. As a machine processes multiple utterances of "cuh-ah-tuh," it begins to connect 'cuh' with 'ah', and 'ah' with 'tuh', eventually linking these sounds to the letters that form the word.

Thus, a machine learns to 'hear' and recognize "cat."

Within just a decade, AI had notably advanced in its ability to interpret human speech—a significant stride for a field that had only been formally named fifteen years earlier. However, despite these advancements, the subsequent redirection of funds towards Cold War efforts signaled the onset of what would be known as AI's 'nuclear winter,' a period marked by reduced enthusiasm and support for artificial intelligence research. This shift underscores a recurring

---

[28] Rabiner, Lawrence, and Biinghwang Juang. "An introduction to hidden Markov models." *ieee assp magazine* 3, no. 1 (1986): 4-16.

theme in AI's history: while technological progress is swift, it is also at the mercy of broader societal and political forces.

Sometimes You Need To Go Back

In the burgeoning field of artificial intelligence, one landmark piece of work emerged from Paul Werbos in his 1974 paper, "Beyond Regression: New Tools for Prediction and Analysis in Behavioral Science," submitted to Harvard University. This paper heralded a pivotal advancement in the development of **artificial neural networks**, which are systems designed to mimic the operational mechanics of the human brain and have since become a cornerstone of AI.

Werbos's research introduced a critical mechanism known as **backpropagation**, a method for training these "artificial brains." Backpropagation is based on the concept of revisiting and adjusting the weights that link various neurons within the network to influence the overall output (remember that from part 1?). To understand backpropagation, consider a new guitar: when you first strum it straight out of its case, it might sound dissonant out of tune. In response, you adjust each string's tuning head and strum again. This iterative process of tuning and retuning, guided by the feedback from each adjustment, mirrors the backpropagation process in neural networks.

In this analogy, the guitar strings represent the neural connections, or "weights," between neurons. These weights are crucial as they determine the strength and influence of one neuron on another, allowing for an almost infinite array of adjustments as the neural network refines its ability to predict and analyze accurately.

As we delve deeper into the 1980s, we'll see that backpropagation shifted from a theoretical concept to

a practical tool, laying the groundwork for teaching
artificial neural networks more effectively.

Werbos's methodology of continually adjusting and improving upon learning processes fundamentally shaped the ways artificial intelligence systems learn, akin to a musician perfecting a melody. This process of constant refinement and adaptation became foundational, not just in AI's approach to problem-solving but in defining how these systems evolve over time to become more accurate and efficient.

The Mansfield Amendment & The Lighthill Report

While the term "AI winter" refers to a period in the 1980s when enthusiasm and funding for artificial intelligence sharply declined, the seeds of this slowdown were planted much earlier, rooted deeply in legislative changes from previous decades. A significant precursor to this chilling effect in AI development can be traced back to a seemingly innocuous line of legislation.

In 1969, Senator Mike Mansfield introduced a pivotal sentence into a Department of Defense authorization bill—a sentence that would have far-reaching implications for the future of AI. This line mandated that the Department of Defense (DoD), and by extension DARPA, which was then the principal financial backer of AI research, could only use its funds for research projects or studies that had "a direct and apparent relationship to a specific military function."

This single directive subtly redirected the trajectory of AI research by significantly narrowing the scope of what could be funded. Prior to this, DARPA had broadly supported a wide range of AI initiatives, fostering innovation with relatively few restrictions. The introduction of this criterion for funding effectively stymied many projects that could not explicitly tie their objectives to direct military applications, leading to a gradual but inevitable reduction in both the scope and the ambition of AI research.

This legislative change was a crucial factor contributing to the onset of the AI winter. It serves as a profound example of how governmental policy, often enacted with immediate practical concerns in mind, can have deep and lasting impacts on the development of cutting-edge technologies, influencing the direction of entire fields for years to come.[29]

In other words – no guns, no funds. This meant AI and its scientists were left out in the financial cold, without the ability to spend where they saw fit for the good of the science.

The onset of AI's "winter" was not solely due to legislative changes in the United States. Across the Atlantic, another significant critique further chilled the climate for artificial intelligence research. This critique was encapsulated in the Lighthill Report, released in 1973 in England, which echoed a similar sentiment to the Mansfield amendment but approached it with a slightly different nuance.

James Lighthill argued that research funding should be tightly coupled with the capability to solve tangible, real-world problems. By his assessment, artificial intelligence did not meet this criterion. Despite notable advancements in the preceding decades—such as solving complex mathematical problems and developing games—AI had, in his view, failed to effectively address practical issues that impacted everyday life.[30] This would be a recurring theme, and one AI would be saddled with until the turn of the millennium.

Moreover, Lighthill highlighted a critical issue within the AI community: overpromising. Researchers, in their enthusiasm, had often promised far more than they had managed to deliver. This discrepancy between expectations and actual achievements made it difficult for the field to justify continued substantial investment,

---

[29] https://pubs.acs.org/doi/pdf/10.1021/ac60289a600
[30] https://www.aiai.ed.ac.uk/events/lighthill1973/lighthill.pdf

particularly from public funds which were increasingly scrutinized for practical returns.

Lighthill's skepticism was influential, casting a long shadow over the field. His report contributed to a growing skepticism about AI's utility and feasibility, further entrenching the funding difficulties initiated by legislative changes like those introduced by Senator Mansfield. Together, these factors significantly dampened the momentum of AI research, leading to a period where funding, interest, and development in artificial intelligence markedly declined.

The 1980's: Winter

"If science teaches us anything, it teaches us to accept our failures, as well as our successes, with quiet dignity and grace."

[starts beating up the creature]

"I'll get you for this!"

--- Dr. Frankenstein, Young Frankenstein

The retirement of Shakey the Robot marked an early harbinger of the approaching AI winter, particularly reflective of the impact of DARPA's shifting funding priorities. Shakey, a pioneer in robotics funded significantly by DARPA, symbolized the golden era of artificial intelligence investment. However, by 1972, as DARPA's support began to wane, so did the momentum behind projects like Shakey.

The introduction of Flakey the Robot in the 1980s, intended as Shakey's successor, did not capture the same level of enthusiasm or impact. Although Flakey achieved a commendable second place at the Association for the Advancement in Artificial Intelligence Robot Competition in 1992, the accolades came too late to maintain the initial excitement.[31]

---

[31] Congdon, Clare Bates, Marc Huber, David Kortenkamp, Ulrich Raschke, Terry Weymouth, and Enrique Ruspini Daniela Mustoz. "CARMEL vs. Flakey: A comparison of two robots." Preprint from the AI Lab o

Unlike Shakey, Flakey incorporated fuzzy logic, a significant leap in AI technology that allowed for decision-making based on probabilities rather than strict yes or no choices. This advancement enabled Flakey to think more flexibly and react to a range of scenarios in a more human-like manner.

It also made him a predecessor to the modern rice cooker, which uses a computer chip to account for human errors like too much water or not enough rice.[32] The first fuzzy logic rice cooker came out in 1983[33], which might make it one of the first 'smart' devices in the modern home.

This adaptation of fuzzy logic from robotics to everyday appliances underscored a pivot in AI applications, moving from laboratory curiosities to practical, everyday utilities. Thus, while Flakey might not have matched Shakey's fame, the technology behind him foreshadowed the broader integration of intelligent systems into the fabric of daily life, heralding the rise of 'smart' devices in modern homes.

As the field of artificial intelligence matured, it became increasingly clear that teaching machines to play games or solve mathematical puzzles was neither boosting the balance sheet nor enhancing U.S. military capabilities. In the era of Ronald Reagan and Margaret Thatcher, amidst the intensifying Cold War, priorities were sharply focused on economic and military strength.

The early AI boom had faded. America was transitioning from the aftermath of one war while escalating its strategic maneuvers in another. Cultural icons like MTV and Diet Coke emerged as Shakey the Robot was retired, and the world of AI had not yet fulfilled the

---

[32] https://dl.acm.org/doi/abs/10.5555/1867270.1867405

[33] https://spectrum.ieee.org/the-consumer-electronics-hall-of-fame-zojirushi-micom-electric-rice-cookerwarmer

David Lloyd

lofty expectations of creating a machine that could emulate a human, leaving many scholars disheartened.

The prospects for artificial intelligence during this period looked bleak. Funding had dried up, and there was a palpable fear among researchers that interest in AI might wane as drastically as the financial support. This period marked AI's first significant downturn, or "flash freeze," where the absence of robust funding like that previously provided by DARPA curtailed further developments in playful or experimental robotics.

This challenging time forced AI to evolve from its academic and experimental nursery into the real world. If the Dartmouth Conference was AI's coming-of-age ceremony, by the 1980s, AI had ostensibly graduated from college, now needing to find a practical and sustainable job.

It was time to put the thinking machines to work, harnessing their capabilities in meaningful, productive ways that could stand up to the socioeconomic demands of the era.

The New Experts

It took almost a decade for the reverberations of the Lighthill report to dampen the enthusiasm around AI. Throughout the 1970s, scientists soldiered on, pushing the boundaries of what was possible despite the looming threat of financial drought. Tools like PROSPECTOR and MYCIN were beacons of hope, illustrating that AI could have practical, profitable applications. By the early 1980s, these expert systems were not just theoretical experiments; they were functional technologies that could potentially boost both economic and military capacities—if they were effectively commercialized.

The dawn of the 1980s saw the emergence of what were called knowledge-based systems—AI applications explicitly designed for commercial viability. One of the pioneering examples was R1/XCON, an expert system developed for Digital Equipment Corporation (DEC). This system directly impacted the company's bottom line by automating the configuration of computers based on customer orders—a task previously performed manually by DEC employees, which was prone to errors and inefficiencies. Before the introduction of R1/XCON, such mistakes were costing DEC up to $40 million per year.[34]

R1/XCON revolutionized this process, significantly reducing the incidence of costly errors and streamlining operations to ensure that systems were delivered and configured correctly. This was one of the first instances where an AI system not only performed a critical business function but also demonstrated a direct return on investment, paving the way for broader acceptance and integration of AI technologies in business operations.

The transformation from experimental curiosity to commercial asset marked a pivotal shift in the AI landscape. As AI began to prove its worth in real-world applications, it slowly started to reclaim the momentum that had been stalled by the skepticism of the late 1970s. This period set the stage for AI's resurgence, as it began to be viewed not just as a scientific endeavor but as a vital component of business and strategic operations.

Indeed, the potential of AI to revolutionize industries was vividly demonstrated by the R1 system, a remarkable piece of technology developed after four years of diligent work by a team largely new to the field of artificial intelligence. R1, later commercialized as XCON, was designed to optimize computer system configurations based on customer orders—a task it performed with singular focus and effectiveness. Unlike systems like MYCIN, which considered

---

[34] Hotz, Lothar, Alexander Felfernig, Andreas Günter, and Juha Tiihonen. "A short history of configuration technologies." Knowledge-based Configuration– From Research to Business Cases (2014): 9-19.

multiple hypotheses or outcomes, R1 was programmed to achieve one specific goal: ensure that each Digital Equipment Corporation (DEC) computer was configured correctly before shipment. This sharp focus not only allowed R1 to match the expertise of human workers but, in many cases, to replace them.[35]

The commercial success of XCON was profound. Integrated into DEC's operations—DEC later became part of Compaq, which was subsequently acquired by Hewlett-Packard—this system helped generate an additional $40 million annually. This significant boost to DEC's bottom line served as a powerful case study for other companies, prompting them to explore similar AI solutions.

The ripple effects of R1's success extended beyond immediate financial gains. It spurred the growth of industries dedicated to supporting such technologies, with companies like Symbolics, LISP Machines, and Texas Instruments leading the way. These firms provided environments where programmers could engage with object-oriented languages that allowed them to define objects, attributes, and behaviors in an intuitive and dynamic way. For instance, a programmer could define a cat with coordinates X and Y, and assign behaviors such as "run to" a specified location.

In these systems, behaviors and interactions between objects were not pre-programmed or "hard coded" by the programmer but emerged from the relationships and attributes defined within the AI. This approach allowed for complex, self-organizing systems where scenarios could evolve naturally based on the parameters set by the AI model. For example, if the coordinates placed the cat on a sun-warmed cushion, the AI could determine that the cat would likely fall asleep; if the coordinates showed a bird in flight, the cat's response would be different.

---

[35] McDermott, John. "R1: A rule-based configurer of computer systems." Artificial intelligence 19, no. 1 (1982): 39-88.

This era marked a significant turning point in AI, showcasing its capability not just in mimicking human tasks but in creating systems that could think, adapt, and interact in ways that were fundamentally transformative. As AI continued to evolve, these early successes underscored the vast potential of intelligent systems to reshape industries and redefine the boundaries of what machines could achieve.

> *Conceptually this thinking is an important shift.*
> *The less you must directly and constantly "program"*
> *a computer or adjust the rules, the better. The whole*
> *goal behind machine learning in AI is for the computer*
> *program to train (or program) itself.*

The burgeoning new industries born from the advancements in artificial intelligence not only revolutionized existing technologies but also spurred the development of innovative software, hardware, and programming languages. One significant language that emerged during this era was PROLOG, an acronym for 'PROgrammation en LOGique.' Originally developed in the early 1970s, PROLOG truly came into its own in the 1980s, aligning perfectly with the expert systems that were gaining prominence during that period.

Like LISP, PROLOG was constructed on a foundation of logic— rules and facts that guided its operations—making it a powerful tool for building AI applications. It became particularly favored in Europe and was integral to Japan's ambitious Fifth Generation Computer Systems program, which aimed to leapfrog existing technologies by developing a supercomputer capable of processing logical **inferences** at high speeds.[36]

> *An Inference is the process of drawing conclusions*
> *or making predictions based on data, algorithms,*

---

[36] Warren, David HD. "A view of the fifth generation and its impact." *AI Magazine* 3, no. 4 (1982): 34-34.

models, or a set of rules or models. So while a model is the mathematical outcome of training a model, inference is the prediction resulting from the model.

Japan's commitment to this initiative, underscored by a substantial $400 million investment, represented not just a bold step forward in technology but also a challenge to American dominance in computer science, particularly the American stronghold on programming languages like LISP. This move by Japan catalyzed a renewed interest and investment in AI within the United States. In response, DARPA reinvigorated its funding for AI research, pouring an additional billion dollars into the development of intelligent machines between 1983 and 1993. This resurgence of support marked a definitive end to the AI winter—a period of reduced funding and interest in artificial intelligence—and heralded a new era of innovation and growth in the field.[37] Mansfield, who?

This renaissance, sometimes playfully referred to as the 'AI spring,' saw not only a revival of funding but also a revitalization of enthusiasm and optimism around AI technologies. The decade following the resurgence was marked by significant advancements that not only caught up with, but in many cases surpassed the innovations happening elsewhere in the world. This period underscored a crucial lesson: competition and challenge often lead to renewed vigor and breakthroughs, pushing industries toward greater achievements. In this narrative, the early setbacks and challenges of AI, represented by figures like Senator Mike Mansfield and his restrictive funding amendments, became mere footnotes in the broader story of AI's triumphant march forward.

---

[37] Strategic Computing: DARPA and the Quest for Machine Intelligence, 1983-1993

Backpropagation Revisted

Despite the setbacks of the early 1980s, the field of artificial intelligence was regaining momentum. One particular challenge during this period stemmed from the response to Marvin Minsky's influential book "Perceptrons." His argument for "frames" to structure information—much like the branches of a maze—had initially pushed discussions on artificial neural networks into the background.

As AI began to find its footing once more, the industry saw a revival of interest in foundational ideas that had been overshadowed. Among these was the concept of backpropagation, developed by Paul Werbos, which was now gaining the recognition and support it deserved.

Backpropagation is predicated on the notion that for neural networks to effectively mimic the human brain, they require multiple layers—far from the simplistic models initially proposed. Werbos used a vivid analogy to explain this need: a human who forgets they are running from a tiger the moment they look away would not survive long, a testament to "natural selection," according to Werbos.[38] Thus, neural networks must incorporate a form of "backward knowledge" that allows them to remember and adjust based on new information or changes over time. This mechanism is crucial for learning from mistakes and improving over subsequent iterations, mirroring the evolutionary adjustments seen in natural organisms.

This period marked a crucial pivot for AI—from theoretical exploration towards practical, robust systems that could learn, adapt, and evolve. The re-emergence of backpropagation was not just a technical advancement but a metaphorical moment in AI's history, signaling a return to complex, multi-layered thinking about how

---

[38] Werbos, Paul J. "Backpropagation through time: what it does and how to do it." *Proceedings of the IEEE* 78, no. 10 (1990): 1550-1560.

David Lloyd

machines learn and interact with the world. This rejuvenation of interest underscored a broader readiness within the AI community to tackle more sophisticated challenges and refine the systems that are now foundational to modern AI applications.

As institutions like MIT and Stanford propelled the advancement of artificial intelligence, the University of Toronto also emerged as a key player in the AI landscape, largely due to the pioneering efforts of researchers such as Geoffrey Hinton, Yoshua Bengio, and Richard Sutton.

Dr. Geoffrey Hinton, often hailed as the "Godfather of Deep Learning", typically including Bengio and Le Cunn, joined the University of Toronto in 1987 and significantly advanced the field of neural network research. Building on the foundational work of Paul Werbos on backpropagation from the previous decade, Hinton's contributions, including the development of Boltzmann machines, were instrumental in shaping the deep learning techniques that are prevalent today. His research not only enhanced the understanding and functionality of neural networks but also set the stage for the transformative applications of deep learning in AI—from speech recognition systems to autonomous vehicles.

The name Boltzmann is from work done by Austian physicist Ludwig Boltzmann based on his work in statistical mechanics. Hinton's machine used stochasticity, or randomness in calculations.

Hinton's influence extended beyond his technical achievements; his presence at the University of Toronto attracted other top talents and helped Canada become a crucial hub in the global AI research community. This concentration of expertise fostered an environment of innovation and collaboration that continues to drive forward the boundaries of what artificial intelligence can achieve.

Additionally, Yoshua Bengio, another key figure in the deep learning community, completed his PhD at McGill University in Montreal completing it in 1991[39], another "Godfather". Richard Sutton, known for his contributions to reinforcement learning, also conducted material research though the same period. Later, he went on to join the faculty at University of Alberta amongst other achievements[40].

> *Reinforcement learning and Reinforcement Learning from Human Feedback are two important approaches to train artificial neural networks through rewards systems both automatically and through human intervention covered in part two.*

Backpropagation is a method that allows an artificial neural network to refine its learning process, akin to perfecting a recipe. Remember our example earlier, baking a cake for the first time? You had only basic instructions and a list of ingredients: milk, flour, sugar, vanilla, salt, and baking powder. You mix these ingredients (inputs) in certain amounts (weights), combine them (layers), and bake at a set temperature (output).

However, if the cake turns out too dry and flat, you'd employ backpropagation: revisiting and adjusting the ingredient measurements based on the outcome, then trying again. With each iteration, the cake improves (perhaps not, which also changes the weights)—first becoming moister, though perhaps still not fluffy. This iterative refining process continues until the cake reaches perfection. This sequence of trial, error, and adjustment is often referred to in AI as a "forward-pass" followed by "backpropagation."

This backpropagation approach underpins the training of neural networks, forming the core of machine learning techniques still

---

[39] https://en.wikipedia.org/wiki/Yoshua_Bengio
[40] https://en.wikipedia.org/wiki/Richard_S._Sutton

prevalent today. These methods allow machines to learn from expert feedback and adapt over time. Yet, during this decade, while these neural network systems were evolving, knowledge systems still heavily relied on human-written instructions. Backpropagation was just starting to transform how machines could learn autonomously.

'Knowledge engineering' relates closely to expert systems but takes it a step further by designing machines to perform tasks usually done by human experts. Unlike earlier AI systems, knowledge engineering applications quickly reached a point of commercial viability. This was achieved through a feedback loop like backpropagation, where machines acquired knowledge from experts, learned to represent this knowledge, and validated their responses. This process converted complex data into actionable if/then statements that the machines could use to perform expert-level tasks. Crucially, in these systems, it was the experts themselves who trained the machine directly; the machine did not learn in isolation.

These developments highlight a significant shift in artificial intelligence from simple programmed responses to dynamic learning systems capable of adapting and improving autonomously. This shift not only enhanced the capabilities of AI systems but also broadened their applicability across various sectors, paving the way for more advanced and autonomous systems.

These expert system shells provided a flexible framework that allowed large organizations and government agencies to tailor the AI to their specific needs without requiring a knowledge expert or researcher to guide the process every step of the way. To implement these systems, an engineer simply needed to consult with a human expert—be it a nuclear engineer, physician, or credit card analyst—to extract their decision-making logic and translate it into rule-based commands.

This approach of converting expert knowledge into a series of logical rules and decisions epitomizes the transition from "If this, then what?" to "That happened, what now?" scenarios. It exemplifies how AI evolved from a theoretical tool into a practical application, fundamentally changing how organizations approached problem-solving and decision-making, streamlining processes, and enhancing accuracy in specialized fields

> In many ways the expert systems of the 70s and 80s can be seen in a similar light to the way in which "prompt" engineering is being used today with Large Language Models like OpenAI (GPT-4), Google (PaLM 2), Anthropic (Claude), Cohere (Various), and Open Source (Llama). Historically experts encoded their knowledge in systems and then those systems posed questions to us to clarify a result. Now, we are using prompts to specifically tell the various Large Language Models what to do and the order to do them in. We have moved from having to dump our "knowledge" into systems to instructing systems how to use their knowledge to surface answers.

One innovative system developed during this era was PUFF, an artificial intelligence designed to assist in diagnosing pulmonary function.[41] Patients would blow into a tube connected to a machine that analyzed various factors such as flow rate, age, and smoking history. PUFF then generated diagnostic reports comparable to those produced by a physician.

PUFF was developed in collaboration with a trained pulmonary physiologist and utilized if/then statements to process data. Remarkably, it performed at an expert level between 90-100% of the time. The paper detailing PUFF's capabilities was released in

---

[41] Feigenbaum, Edward A. "Themes and case studies of knowledge engineering." *Expert systems in the micro-electronic age* (1979): 3-25.

1979, marking a significant milestone in the field of knowledge engineering, which continued to evolve and improve throughout the 1980s until it reached a stage ripe for commercialization.

The pathway to commercializing these AI systems involved what were known as Knowledge-Based Expert Systems Shells. These shells were essentially expert systems stripped of their specific domain knowledge—think of MYCIN without the medical expertise or PUFF without its specialized pulmonary function data. Programmed in languages like LISP, these shells contained the essential if/then logic necessary to build decision trees but were devoid of any specific content, making them highly adaptable.

If this, then what? That happened, what now?[42]

The first large scale commercial implementation of these knowledge system shells was at American Express. Amex cardholders' credit card transactions were approved or denied using these if/then statements to gauge potential for fraudulent use of the card. If the card had been recently reported lost, then the AI was taught to deny the transaction, and so on. It was unique in that it could operate both on its own (ie. denying a large transaction on a lost card immediately) and in conjunction with an authorizer (i.e. creating a summary screen of the customer's credit to save the authorizer time).

Amex's knowledge system rollout was a perfect example of the commercialization of early AI. The code was written by Inference Corporation using the knowledge shell ART (which stood for, Automated Reasoning Tool). The software was distributed by Symbolis, and the mainframes were from IBM: three industry juggernauts just picking up speed.

The entire Amex AI ecosystem was created in the span of 3 years. From 1985 to the rollout in 1988, Amex not only created a viable AI

---

[42] https://developer.att.com/blog/the-evolution-of-ai

tool of its own, it had the software, hardware, and programming to support it.[43] This meant the creation of business, jobs, and commercial independence in the field. Spring rolled into summer, and AI fell into step with another economic boom.

As neural networks and backpropagation turned into commercially viable approaches, computer hardware and software had the financial incentive to keep up. DARPA funding would re-emerge, but it was no longer the only source of income spurring the field along. AI was now a booming market, but arguably not an overt one. It wasn't packaged in AI paper, but AI was slowly but surely working its way into different aspects of our lives.

DARPA's involvement was no small factor in the change. Even as AI headed into another 'winter' at the end of the decade, advancements were continuing.

Robots

In the whirlwind of the 1980s, robotics and artificial intelligence danced a tandem that propelled both fields into the commercial spotlight. As neural networks began redefining technology, robotics wasn't far behind, capturing the imagination and investment of the era. The PUMA robot systems (Programmable Universal Machine for Assembly) emerged as pivotal players, designed to revolutionize industrial machinery in the way that neural networks were transforming software.

Stanford and MIT, pioneers in academic innovation, were among the first to establish dedicated robotics laboratories. The brainchild of Stanford graduate student Vic Scheinman, the PUMA robot was developed during his time at Vicarm, a company he founded that was later acquired by Unimation in 1977. Unimation, a behemoth in

---

[43] Stark, Marilyn. "Authorizer's Assistant: a knowledge-based system for credit authorization." In *Wescon/96*, pp. 473-477. IEEE, 1996.

industrial robotics since the 1960s, it became the arena where MIT engineer Mitch Weiss refined the PUMA technology. In these halls, the blending of academic brilliance and practical application took shape, leading to significant advancements.

The PUMA robots, much like their counterpart, the SCARA—developed by Scheinman's Stanford lab mates and commercialized by Adept—were not only fast, operating at three times the speed of a human worker, but also programmable. This ability to learn and perform tasks autonomously made them groundbreaking. Yet, despite their potential, there was hesitance among factory workers to embrace these machines, a reluctance that ultimately saw Unimation's decline and selling off its assets.

Nevertheless, the story of PUMA and its ilk is not just about the fleeting AI boom of the 1980s; it's about laying the groundwork for the symbiosis of commercial and computational success that would define future decades. This tale, marked by human ingenuity and the relentless pace of progress, sets the stage for understanding the transformative power of robotics and AI in our lives today.

The 1990's: Bigger Data, Small Machines

"It's fair to say what's going on today is like the arrival of the printing press, or the telephone or the radio. And these communications tools did have pervasive effects. They made the world a smaller place."

--- Bill Gates, 1996

Going back to dinosaur lore, John Hammond's Jurassic future depended on finding the missing link. His dino sequence had some gaps, and to fill them, he substituted the next best thing: frogs. Introducing frogs into the dinosaur-mosquito gene pool had unintended consequences, as human creations often do. The genome splicing that should have resulted in only female dinosaurs did not. Once again, life broke free. It found a way.

In the 1990s, one of the most significant contributors to artificial intelligence was not a person or a breakthrough technology, but something far more ubiquitous: our data. As knowledge-based systems evolved, so too did the commercialization of AI. Meanwhile, geopolitical climates shifted dramatically—the Cold War thawed, and East and West Berliners dismantled the wall. With the end of Cold War defense spending in sight, AI needed to carve out new pathways to sustainability.

And initially, it succeeded—mostly.

By the late 1980s, expert systems had infiltrated nearly every major field. While not yet embodying true AI, these systems nudged the field toward tangible commercial applications that mimicked intelligent behavior. Personal computers, now more powerful and accessible than ever, were no longer confined to university labs or elite research facilities.

The ecosystem surrounding AI also burgeoned, with industries related to software, hardware, and programming tools sprouting globally. The expertise of specialists became encapsulated within expert systems shells, accessible in fields ranging from medical diagnostics to finance, geology, and circuit design.

Within just three decades, the world had not only seen a surge in population and disposable income but also an explosion in both demand and supply. Consumerism adopted a new universal language, more pervasive than programming languages like PROLOG or LISP. Identity was now defined by cars and clothes, with every manufactured item—from goods to services—crafted to maximize satisfaction, intensify pleasure, and justify expense.[44] Both the growing amount of data as the internet more formally became available in 1993 and the rise in consumerism would be key ingredients in driving AI.

Yet, such rapid growth fostered caution among experts. Marvin Minsky had learned from past overestimations that hyping AI's potential could lead to disappointment if expectations were not met.[45] The complexities of the human mind, particularly in handling multiple decision-making variables, were just beginning to be understood. At the American Association for Artificial Intelligence's 1984 meeting, Minsky and Roger Schank warned of another impending AI winter, precipitated by unsustainable hype. They were right; the downturn arrived in 1987.

AI had become tightly linked with the global economy. Even without the post-Cold War cutbacks from the Department of Defense and the fallout from Japan's Fifth Generation Computer Project, AI was still subject to the whims of the market, demonstrating that the

---

[44] Cross, Gary S. An all-consuming century: Why commercialism won in modern America. Columbia University Press, 2000.
[45] Wright, Robert. "Thinking Machines." The Wilson Quarterly (1976-) 8, no. 5 (1984): 72–83. http://www.jstor.org/stable/40257629.

fate of this burgeoning field was as much tied to global economic trends as it was to technological advancements.[46]

At the turn of the decade, artificial intelligence found itself in a precarious position—like a Silicon Valley start-up suddenly bereft of its venture capital, AI had bills to pay but seemingly no means to do so. The formidable computer systems that had fueled the previous boom had become relics of the past. The LISP machines, developed by Symbolics and once the backbone of AI's computational power, were phased out. Technology was advancing, becoming sleeker, more user-friendly, and crucially, more affordable. IBM and Sun Microsystems led the charge, offering cheaper machines that could run LISP and other programs without the need for specialized hardware.

Expert systems, a pivotal component of AI's toolkit, were reimagined for this new era. These systems, along with other AI elements, began to harness the power of rule-based 'if/then' statements combined with case-based reasoning and fuzzy logic. This fusion allowed AI applications to make inferences and recognize partial truths, simulating a type of 'memory' that could apply learned cases to new situations. The evolution was emblematic of a broader shift in the world of technology a move from monolithic, specialized machines to versatile, accessible platforms that promised to democratize the power of AI, embedding it more deeply into the fabric of everyday technology.

Before the advent of artificial intelligence, human experts—like physicians and lawyers—relied on the analysis of previous cases to inform their current decisions. This method, integral to professions requiring judgment based on precedent, provided a foundation for what would become a key strategy in AI, **Case-Based Reasoning**

---

[46] https://www.washingtonpost.com/archive/business/1992/06/02/japanese-government-ends-development-of-computer/97184ba9-7177-4516-82f9-d684aecd1f24/

David Lloyd

(CBR). In this system, solving a problem or making a diagnosis benefited immensely from referencing earlier examples.

Historically, expert systems focused on singular problems—diagnosing an illness or detecting credit card fraud. However, with the adoption of case-based reasoning, these systems evolved into robust machines capable of handling diverse tasks.

Kolodner explains, "We are more competent the second time because we remember our mistakes and go out of our way to avoid them." The effectiveness of a case-based reasoning system hinges on four key factors:

- The experiences it has had,
- Its ability to interpret new situations through the experiences,
- Its skill in adapting, and
- Its capability to evaluate and learn."[47]

Provided it can recall a relevant case when needed, adapt it to a new context, and learn from each experience, such an AI can indeed manage an array of challenges.

All it needed now was the data.

Big Data Part 1: Deep Learning & Network Architecture

The 1990s marked a pivotal era in artificial intelligence, particularly in the way AI systems came to represent knowledge. To understand knowledge representation, imagine trying to solve a problem: your internal thought process creates a mental model of the problem, which is inherently external. For instance, envisioning a circuit model doesn't

---

[47] Kolodner, Janet L. "An introduction to case-based reasoning." *Artificial intelligence review* 6, no. 1 (1992): 3-34.

illuminate a bulb; it's the act of wiring it based on that mental blueprint that does.

For machines, this means translating human problem-solving techniques, critical thinking, and what we often call 'common sense' into a language they can understand. This translation occurs through various forms—semantic networks, neural nets, frame languages, and deep learning—which we covered in part one. Initially, the focus of the 1990s was less on the sheer volume of data (which by today's standards was minuscule) and more on refining the structures or architectures of AI technologies. These architectures determined how data was stored and made accessible for AI systems to learn and function.

The decade also saw advancements in neural networks, building upon previous concepts like backpropagation—a method for fine-tuning the weights (remember guitar tuning) and biases in a neural network. This period reintroduced and enhanced backpropagation through the development of **Convolutional Neural Networks** (CNNs), a significant breakthrough which we explored. CNNs revolutionized how machines interpret images, setting the stage for the sophisticated image recognition technologies we use today. In this way, the 1990s did not just expand AI's data capabilities; they deepened its very understanding of the world.

Rise of Le Machines

In 1989, Yann LeCun and his colleagues introduced a groundbreaking method using backpropagation to learn postal codes, fundamentally changing how computers perceive and interpret images. Before this innovation, machines recognized images through vectors—digital graphics defined by mathematical equations. Essentially, this method involved giving a computer a set of drawing instructions and having it piece the image together.

Vector images, scalable and versatile, are ideal for logos and digital graphics and have numerous AI applications. For instance, the coordinates of a vector might be used in algorithms like k-nearest neighbor for predictions or serve as 'templates' for AI to compare against real-world images. However, these methods demanded considerable expertise and effort.

LeCun opted for a simpler, more intuitive approach. Instead of complex vector image, he fed the machine straightforward pixel images. Remember the work of Dinneen and Selfridge at Lincoln Lab? In a similar vein, LeCun normalized handwritten numbers for size and applied a layered approach using backpropagation to analyze their features. He dubbed his creation LeNet, turning the concept of 'deep learning' into a tangible reality.

Imagine a pixel as a single dot in a 100 x 100-pixel image, comprising 10,000 pixels in total. Picture a number (0-9) as a puzzle broken into pieces—these pieces are the features the model needs to recognize. In LeCun's model, these puzzle pieces are distributed across various layers, each tasked with a specific function. For example, some layers, like the edges of a puzzle, are sorted out first to simplify the subsequent steps.

Each layer in LeNet uses backpropagation, akin to adjusting our cake recipe in baking. Suppose a cake needs less sugar; you'd tweak the recipe and bake it again until it's just right. Similarly, LeCun's layers learn from each backward adjustment to enhance accuracy. Some layers, known as convolutional layers, extract specific features like the curve of a letter 'o'. Others, called pooling layers or down-sampling layers, simplify the image to make it easier to process, akin to how different brain regions specialize in processing various sensory inputs before integrating them.

As each layer is resolved, like completing portions of a puzzle, the picture becomes clearer. You might start to guess the image even before all pieces are in place. In LeCun's model, each data point is a 'node' in the neural network, akin to the machine's thought process. With each example it processes, the machine learns which layers are crucial, gradually improving its ability to accurately identify the image, whether it's a dog, a cat, or a letter 'o'. This method not only democratized machine learning but also illustrated the potential of neural networks in practical applications, heralding a new era in AI.[48]

Recurrent Neural Networks (RNNs) represent a fascinating branch of deep learning tailored for analyzing sequential data, such as daily stock market trends or the arrangement of words in a sentence. What sets RNNs apart is their ability to remember prior data, making them crucial for tasks where historical context is key. Initially, their widespread use was limited due to their size and complexity, but they certainly made their mark.

A significant breakthrough came with the development of **Long Short-Term Memory** (LSTM) networks, which addressed a critical issue in backpropagation related to memory gaps. For instance, consider the sentence, "I was born in Spain. I was raised there, too. I still speak fluent..." The answer, "Spanish," though logically linked, is separated from the clue "Spain" by several words—a challenging gap for machines to bridge. LSTMs manage this by introducing 'gates' that regulate the flow of information, allowing the system to maintain or discard data as needed.

To visualize the function of LSTMs, imagine you're engrossed in a lengthy fantasy novel series. Over time, as the saga unfolds through numerous volumes, it's natural to forget earlier details essential for understanding future plot twists. Here, you might take notes to track key events, but excessive notetaking can detract from the pleasure of

---

[48] LeCun, Yann, Bernhard Boser, John Denker, Donnie Henderson, Richard Howard, Wayne Hubbard, and Lawrence Jackel. "Handwritten digit recognition with a back-propagation network." *Advances in neural information processing systems* 2 (1989).

reading. This is analogous to how LSTM 'gates' operate. Input gates decide which data to store and which to let go, while output gates help transfer essential information to long-term memory, ensuring you're prepared for upcoming surprises and can discuss the plot twists with friends confidently.

In essence, LSTMs enhance a machine's ability to process and remember information over extended sequences, much like a dedicated reader keeping tabs on the sprawling narrative of a complex novel. This ability makes RNNs incredibly powerful tools for applications where understanding the past is crucial to interpreting the present.

It is a very important point to also remember in 1993 we had the formal introduction of the world wide web. While we will talk about the increasing rise of data, the internet not only provided the ability to create massive amounts of content but make it accessible in digital format for easy consumption.

Support Vector Machines

Support Vector Machines (SVMs) arrived on the scene in a pivotal 1995 paper by Corinna Cortes and Vladimir Vapnik. SVMs were adept at drawing lines—metaphorically speaking—which means they excelled at classifying data and performing regression analysis. Essentially, SVMs could sift through a mass of data, such as race times or test scores, and identify the optimal line to distinguish the lower half from the upper half of a dataset.

While SVMs aren't a direct offshoot of the deep learning that typified AI in the '90s AI, SVMs[49] significantly bolstered the machine learning toolkit. They provided new methods for handling data, including imperfect or ambiguous datasets, thereby complementing other AI technologies.

---

[49] Cortes, Corinna, and Vladimir Vapnik. "Support-vector networks." *Machine learning* 20 (1995): 273-297.

Deep learning architectures laid the groundwork for the predictive analytics and machine learning models that are commonplace today.

The 1990s were a complex period teeming with technological advancements. It was a time when AI was quietly burgeoning beneath the surface, much like the subcultures of the Spice Girls and grunge rock. This era didn't just bring catchy tunes and flannel shirts; it also saw a rise in data mining, data collection, and knowledge discovery.

> *It is a very important point to also remember in 1993 we had the formal introduction of the world wide web. While we will talk about the increasing rise of data, the internet not only provided the ability to create massive amounts of content but make it accessible in digital format for easy consumption.*

Predictions were becoming a lucrative business, and AI's ability to make these predictions was growing increasingly precise. Despite the carryover of the 'AI winter' from the late 1980s into the early '90s, it was the advent of big data—and the proliferation of smaller, more powerful machines—that would eventually catalyze a significant shift. As AI systems were trained on increasingly larger datasets, they became more embedded in our everyday lives. The technology advanced, positioning AI to predict vast swathes of consumer behavior over the next two decades.

At this juncture, the focus was on exploring the possibilities of what could be done with AI, not necessarily what should be done. This mindset set the stage for a period of rapid innovation and, perhaps, a need for introspection about the implications of these advancements.

Big Data Part II: Language and Information

In the 1960s, the initial ventures into processing language via machines began, setting the stage for the burgeoning field of modern artificial intelligence. By the 1990s, the pioneers of AI were exploiting these early technologies to fuel systems like initial search engines of AltaVista and Yahoo, the precursors to the omnipresent Google search. Natural Language Processing (NLP) had evolved significantly, employing a wide range of techniques that remain foundational today.

Consider **n-grams**, for example, somewhat analogous to today's hashtags. In this system, a phrase like 'I like red berries' would be broken down into segments or bigrams—'I like,' 'like red,' and 'red berries'—providing clues about the sentence's structure and content. Statistical language models of the era could even predict text probabilities. For instance, the phrase "It's getting late, time to go to…" the model might suggest 'bed' as the most likely conclusion.

As the decade progressed, these tools became more visible and prolific. Carnegie Mellon scientists demonstrated their prowess by navigating across America in a robotic car. The dot-com boom reached its zenith in the late '90s, and the world witnessed one of the most public displays of AI capabilities with IBM's chess-playing computer, Deep Blue. In 1996, the world chess champion Gary Kasparov triumphed over Deep Blue, holding the line for humanity.

However, by 1997, the tables had turned; Kasparov lost, sparking some of the first real public concerns over the power of artificial intelligence. Despite this, it was hard not to feel as though the future had already arrived. Yet experts like Marvin Minsky remained skeptical. In 1998, shortly after Deep Blue's victory, Minsky highlighted ongoing challenges: researchers were still struggling with to enable machines to grasp real-world problems and common sense.

"A.I. was able to produce all kinds of wonderful things," he remarked, citing programs that outperformed average stockbrokers or repaired equipment. But, he noted, the systems faltered with anything contextual. He recounted an anecdote about an M.I.T. student's AI program faced with a scenario where someone's daughter was kidnapped by the Mafia, and a ransom was demanded. When asked, "What should we do?" the program faltered, ultimately querying, "Why would he pay MONEY to get his daughter back?" This illustrated the stark limitations of early AI in understanding human contexts, underscoring the complexity and unpredictability of real-world interactions—an area where AI still had much to learn.[50]

Ultimately, IBM dismantled Deep Blue. But as the decade (and the millennium) changed, many started to wonder whether 'artificial intelligences' would start striking back. While this wasn't on the menu – and still remains 'off the table' as far as humanity is concerned – the machines were getting smarter, even as the real world started to collapse.

---

[50] https://www.nytimes.com/1998/07/28/science/conversation-with-dr-marvin-minsky-why-isn-t-artificial-intelligence-more-like.html

The New Millenium: AI Goes Mainstream

"From 2005 to 2020, the digital universe will grow by a factor of 300, from 130 exabytes to 40,000 exabytes, or 40 trillion gigabytes (more than 5,200 gigabytes for every man, woman, and child in 2020). From now until 2020, the digital universe will about double every two years."

--- IDC's Digital Universe Study (2012)

"It's not artificial intelligence I'm worried about, it's human stupidity."

--- Neil Jacobstein

As we reached the threshold of the new millennium, the field of artificial intelligence had already endured two significant setbacks, often referred to as "AI winters." Yet, it emerged revitalized, poised to tackle and shape an emerging new world. This era saw AI not just survive but thrive—transformed, commodified, and monetized, broken down into fundamental components and structures. As pioneers of technology ventured into this uncharted terrain, the rest of us could only watch and wait.

Meanwhile, the world grappled with broader economic uncertainties. The dawn of the "internet age" and the onset of the "new economy" it fostered ushered in a period of extreme fluctuations for entrepreneurs and investors alike, characterized by alternating currents of greed and hubris. The zenith of 1999 gave way to a sharp downturn in 2000 and 2001, raising concerns about whether AI might succumb to the economic collapse. Having weathered two AI winters, the question loomed: Could this crash trigger a third, or was AI now robust enough to stand independently?

The New Machines

In the fledgling days of artificial intelligence, financial support came from grants. Alan Turing's sabbatical yielded the concept of Intelligent Machines, the Rockefeller Grant brought the Dartmouth Conference to life, and DARPA's funding sustained AI's early development. By the close of the 20th century, AI had transformed into an industry flush with its own capital. The early success of systems like R1/XCON, which promised $40 million in annual savings, drew the financial sector's eyes to the burgeoning potential of AI.

As the world teetered on the brink of a new era, the "new economy" was still finding its footing. The dot-com boom of the late 1990s was poised to give rise to the software industry and usher in the "internet age," but even the AI vanguards felt the uncertainty of the times.[51]

Yet, it was during this period that significant strides were made in deep learning and neural networks—often thought of as 'the mind' of AI. Technologies like neural networks and Support Vector Machines, refined through the late 1990s, began advancing the frontiers of pattern recognition. Search technologies evolved rapidly; Google, Alta Vista, Bing, and others pushed the boundaries of natural language processing and understanding. The internet, burgeoning with content, acted as fuel to this fire, and our personal data served as the accelerant, igniting a spectacular blaze.

---

[51] Buenstorf, Guido, and Dirk Fornahl. "B2C—bubble to cluster: the dot-com boom, spin-off entrepreneurship, and regional agglomeration." *Journal of evolutionary Economics* 19, no. 3 (2009): 349-378.

David Lloyd

This data explosion provided AI with a rich vein to mine. Recommendation engines on platforms like Netflix honed their ability to suggest content based on user browsing habits. Amazon and other retailers used similar technologies to suggest products, leveraging insights like those from K-nearest neighbor algorithms to predict consumer behavior—sometimes finding correlations as curious as the link between beer and diaper purchases, whether accurate or not.[52]

By the early 2000s, it was clear: AI was not just a scientific endeavor but a commercial powerhouse. The industry was moving away from its reliance on DARPA grants and into a phase where commercialization of data spurred further growth. Predicting consumer behavior had become not just feasible but lucrative, drawing investors and fueling rapid advancements in AI.

This remarkable progress expanded into fields like pattern recognition, computer vision, and speech and language processing. As the new decade dawned, AI was not merely keeping pace with technological advances but often outstripping them, heralding a future where it was faster, stronger, and more integrated into every facet of our digital lives.

Even Bigger Data (Forest and Trees)

Even as the world was still navigating the digital landscape with Windows XP, the early 2000s were all about data—vast quantities of it. In 2006, mathematician Clive Humby famously likened data to "the new oil." This comparison, often misunderstood, wasn't just about profitability. Humby emphasized that, like crude oil, data must be refined to be useful.

---

[52] https://tdwi.org/articles/2016/11/15/beer-and-diapers-impossible-correlation.aspx

Imagine a decision tree. You don't need a background in statistics to grasp its purpose—it's a tool for selecting the best outcome from a series of choices, whether solving a complex equation or deciding whether to eat last night's leftovers.

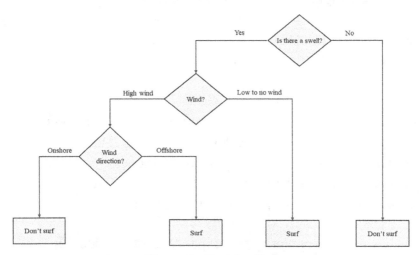

Figure 9 - Decision Tree

Decision trees lay out the steps in making a choice, much like how a computer processes data to reach a decision. For instance, here is a decision tree to decide whether to go surfing, encapsulating both logical and seemingly illogical human decision-making processes.

Marvin Minsky's concept of "frames" was designed to mimic this human decision-making. In a similar vein, Leo Breiman introduced **"random forests"** in 2001. If a decision tree represents a single decision path, a random forest consists of many such paths.[53] Each tree in the forest uses a portion of the data to make predictions, which are then combined to reach a consensus. For example, when presented with a picture of a bear, 75 trees might identify it as an animal, while 25 might mistake it for a shaggy rock. The minority is overruled, and the majority classification prevails.

---

[53] Breiman, Leo. "Random forests." *Machine learning* 45 (2001): 5-32.

This system turns complex human thought processes into something a computer can manage. Random forests expand this capability dramatically. Consider thousands of shoppers at a grocery store on a Saturday, each transaction captured through loyalty cards. The AI sorts through this data like a bear through garbage, ultimately producing insights through a collective judgment of many decision trees. If it finds that shoppers spend more on snacks at the checkout before lunch, the store manager might stock more chocolate bars as a promotional tactic.

AI had evolved beyond needing the lifeline of government grants like those from DARPA; it was generating its own revenue, mastering the handling of extensive data about decisions, predictions, and behavior. Its applications in the business world were becoming indispensable, powered by its ability to turn vast data into actionable intelligence.

The Semantic Web

In May of 2001, just before some pivotal events of the 21st century began to unfold, Scientific American published an article titled 'the semantic web'. Its byline was prophetic: "a new form of Web content that is meaningful to computers will unleash a revolution of new possibilities."[54] At that moment, the world was still reveling in the digital euphoria of Napster downloads and Winamp visualizations, even as the New York Times was on the cusp of declaring the era of 'dot com' as 'dot gone'. [55] [56] The article resonated widely, capturing the attention and imagination of its readers.[57]

---

[54] Berners-Lee, Tim, James Hendler, and Ora Lassila. "The semantic web." *Scientific american* 284, no. 5 (2001): 34-43.
[55] https://www.nytimes.com/2018/10/08/style/dot-com-crash-of-2000.html
[56] https://winampheritage.com/visualizations/Visualizations-1
[57] https://dl.acm.org/doi/pdf/10.1145/3397512

The semantic web was designed to enable machines to understand the relationships between data points. Using tools like **Extensible Markup Language** (XML), which creates hidden labels that help machines read web pages, and **Resource Description Framework** (RDF), which organizes data into triples of things, properties, and values—for instance, the band 'The Cure' is linked to the song 'Pictures of You' by the property 'is the artist behind'. Ontologies weave these pieces together into a coherent logic: bands produce music; thus, 'The Cure' makes music, which can be played on Winamp.

Before the advent of the semantic web, the internet was a vast sea of data, rich in information but lacking meaningful connections. The semantic web bridged these gaps, enabling not just more intuitive data cataloging but also smoother online transactions like automatic reservation confirmations and streamlined AI-driven online shopping experiences. As Tim Berners-Lee[58] had predicted in *Scientific American*, the semantic web fundamentally changed how we conduct business and manage our daily lives, paving the way for a future where data connectivity is seamlessly integrated into every aspect of the digital experience.

Reinforcement Learning

The emergence of the semantic web marked a paradigm shift, enabling 'agents'—specialized software programs—to interact with and interpret data in revolutionary ways. This newfound accessibility spurred AI experts to refine how these tools accessed and learned from information, turning to the well-established principle of reinforcement.

Consider the analogy of a lab mouse: it learns to press a button to receive a sugar cube. Press the correct button and get a reward; press the wrong one and face a mild reprimand. Over time, the mouse

---

[58] https://en.wikipedia.org/wiki/Tim_Berners-Lee

adapts, pressing the correct button more frequently to maximize its sugar intake.

This concept of learning through trial and error underpins the early research into neural networks by pioneers like Yann LeCun, setting the stage for advancements in reinforcement learning. By 2005, Martin Riedmiller was pushing these ideas further with his development of the 'fitted Q' technique, a method that adapts neural networks for more complex decision-making scenarios.[59] To illustrate, remember how we baked a cake through backpropagation. It starts with the outcome—a cake that receives an average rating from human tasters—and works backward, adjusting ingredients like milk and sugar to improve the result. The 'fitted Q' comes into play when the recipe variations are too numerous to test each one individually. Instead, each adjustment receives a 'score' based on its effectiveness, allowing the machine to make educated guesses about which combinations of ingredients might yield the best results. It's akin to a mouse using a mini scorecard to track which buttons yield sugar and which don't, thereby optimizing its actions to secure the maximum reward.

Through these mechanisms, AI does not merely learn; it adapts systematically, turning raw data and past experiences into a finely tuned strategy for success.

From Forest to Mines

In 2006, a groundbreaking textbook emerged from the collaboration of University of Illinois professor Jiawei Han, Simon Fraser University professor Jian Pei, and Micheline Kamber, marking

---

[59] Riedmiller, Martin. "Neural fitted Q iteration–first experiences with a data efficient neural reinforcement learning method." In *Machine Learning: ECML 2005: 16th European Conference on Machine Learning, Porto, Portugal, October 3-7, 2005. Proceedings 16*, pp. 317-328. Springer Berlin Heidelberg, 2005.

the maturation of data mining into a distinct field of research. Although the roots of data mining began to spread in the 1990s, it was not until the early 21st century that it fully blossomed into its own area of scholarly pursuit.[60]

This growth was fueled by an unprecedented explosion of data. Thanks to advancements like the semantic web, which structured internet content, and the burgeoning of online shopping, companies were suddenly flush with information on everything from consumer purchasing habits to personal entertainment preferences. As Han and Pei insightfully noted, necessity is indeed the mother of invention. With such an abundance of data available, it was prime time for AI to roll up its sleeves and delve into this rich soil.

> *As noted in the 90s decade, this continual growth in big data is critical to the speed of AI development in practical terms. As noted above, the semantic web allowed a structured way to describe internet content (helping AI know what the content was).*

The '90s had already hinted at the pivotal role of big data in accelerating AI development, but now businesses across the globe were building colossal models of consumer behavior, incorporating everything from sales transactions to product descriptions and remote sensing data. The volume of data seemed boundless, and the field of data mining exploded with activity, attracting scores of fortune-seekers, much like the gold rushes of old.

> *To put the growth of data in perspective, in 2010 it was estimated that there were about 2 zettabytes of data worldwide[61], it is estimated that by 2025 there will be 181 zettabytes. So how much data is that?*

---

[60] Han, Jiawei, Micheline Kamber, and Data Mining. "Concepts and techniques." *Morgan kaufmann* 340 (2006): 94104-3205.
[61] https://www.statista.com/statistics/871513/worldwide-data-created/

Some round numbers: imagine that a 400-page novel contains about 100,000+ words or 1 megabyte of data to simplify. Then consider the worlds total population is about 8 billion. Once you calculate how many books 181 zettabytes is, divided by the population, each person would have over ~22 million books on their summer reading list in 2025.

Han and Pei emphasized the urgent need for "powerful and versatile tools" to sift through this immense data trove and transform it into structured, actionable knowledge. This necessity has spurred remarkable innovations in data mining, which in turn has solidified and expanded the foundations of AI research. The potential to predict future trends and behaviors from vast datasets meant that AI could self-finance, driven by the lucrative potential of its applications.

This shift has removed the shackles from AI, enabling limitless exploration and application in virtually every domain imaginable. As AI continues to evolve, its capabilities seem only bound by the quantity and quality of data it can analyze..

Big Data Includes Big Images

The domain that saw remarkable growth from the pioneering work of Dr. Fei-Fei Li is deep learning in the realm of images[62]. In 2006, Dr. Li embarked on developing a crucial foundation for computer vision. Her project, which spanned until 2009, culminated in a massive database named ImageNet, containing over 14 million images sorted into 20,000 categories.

ImageNet became a cornerstone for the annual ImageNet Large Scale Visual Recognition Challenge (ILSVRC) starting in 2010. This competition drew global researchers who tested the prowess of their deep learning models in accurately classifying and recognizing

---

[62] https://profiles.stanford.edu/fei-fei-li

images from the ImageNet database. The challenge spurred significant advancements in convolutional neural networks, a topic we highlighted in part one.

Dr. Li's career has been nothing short of illustrious. She serves as Co-Director of Stanford's Human-Centered AI Institute and has directed Stanford's AI Lab. Her influence extended beyond academia when she took roles as Vice President at Google and Chief Scientist of AI/ML at Google Cloud, further impacting the field.

As we stepped into the new millennium, the pace of digital information growth was nothing short of explosive. Data is the bedrock of AI, but as we'll explore in the coming years, two other critical elements emerged to complete the trifecta necessary to propel AI to where it stands today.

The Tens / The Twenties / A New Spring

> *"The new spring in AI is the most significant development in computing in my lifetime. Every month, there are stunning new applications and transformative new techniques. But such powerful tools also bring with them new questions and responsibilities."*
>
> *--- Sergey Brin*

> *"AI is not a decision, it's an evolution."*
>
> *--- Nathan, Ex Machina*

AI had indeed traveled a long distance from its humble beginnings at Dartmouth. Now, it was flush with funding, equipped with advanced tools, and supported by an army of experts. More crucially, it was riding a wave of global public interest.

Artificial intelligence had become a tangible force. We had breathed life into it, and it was beginning to take on a life of its own.

The early 2000s ushered in unprecedented access to data and novel ways to utilize it. AI systems had evolved, mastering data mining techniques and learning to mimic human decision-making processes. Perhaps most significantly, AI had crafted a lucrative ecosystem, where the data it harvested could be processed and interpreted to enhance sales across various industries—consumer goods, food services, restaurants, social networks, clubs, delivery services, and navigation tools. The potential commercial applications of AI seemed boundless, as were the organizations eager to leverage them.

And so was the data.

As customers worldwide embraced these new innovations and tools, the data they generated poured into the burgeoning AI economy, which eagerly consumed this fresh information. For a time, it seemed like a mutually beneficial arrangement: customers willingly provided their data, and companies fed it into the ever-growing appetite of AI systems.

However, it wasn't long before consumers began to notice something peculiar. Mention "Chevy trucks" near a smartphone, and suddenly, Google would bombard that device with ads for the latest Chevy Colorado, a model noted for its robust 35-inch tires and bold attitude. This realization about data privacy and targeted advertising would become a significant concern, but at that moment, the global market was more than content to summon an Uber to explore a new restaurant discovered through Google, take a few pictures with their smartphones, and then drive away in the Chevy Colorado they suddenly felt an inexplicable urge to purchase. Life, it seemed, was good.

Well, sort of…

Deep Learning Revolution

Amid a resurgence of public interest—perhaps spurred by the allure of that Chevy Colorado—researchers perfected deep learning techniques, positioning AI at a fascinating junction of abundant tools and funding. It felt almost like a revival of the early days when robots were first taught to play checkers or solve math word problems. The world had already marveled at IBM's Watson domination on *Jeopardy!*, a feat achieved through a blend of natural language programming, hypothesis evaluation, and dynamic learning. Here was AI, once again, dazzling us by outsmarting and outpacing human competitors.[63]

Then came 2016, a landmark year when Google DeepMind's AlphaGo defeated the world's top Go player, a demonstration of decision-making trees in sophisticated action. Go, while being one of

---

[63] High, Rob. "The era of cognitive systems: An inside look at IBM Watson and how it works." *IBM Corporation, Redbooks* 1 (2012): 16.

the most played games globally, presents one of the greatest challenges for AI due to its vast array of possible moves and the complexity of strategic decision-making. Traditional AI tools struggled to navigate this complexity, but AlphaGo broke through these barriers using neural networks—a blend of human oversight and reinforcement learning that trained the machine in unprecedented ways. The result was astounding: AlphaGo boasted a 99.8% win rate against other programs and swept a human professional 5-0 in a series previously thought to be at least a decade away from possibility. To grasp the complexity, consider that AlphaGo can execute more unique moves than there are atoms in the universe. This sequence of events not only showcased AI's capabilities but also redefined what we thought possible, pushing the boundaries of both technology and the games it could master.

One year after the remarkable success of AlphaGo, researchers at the University of Alberta introduced DeepStack—an AI-powered poker machine that not only mastered gameplay but adeptly managed the uncertainties inherent in poker.[64] DeepStack, a testament to AI's evolution since the 1950s, operates on a standard gaming laptop and makes decisions within three seconds.

The foundations for these advanced tools were laid earlier in the century, accelerated by the influx of funding and public interest. Deep Neural Networks (DNN) evolved from earlier models such as Yann LeCun's LeNet, which transformed handwritten numbers into recognizable digital features. Over time, DNNs expanded to include more layers, allowing the networks to handle more complex features and data, enhancing their accuracy and depth.

By 2012, advancements like 'AlexNet', which utilized eight layers and Rectified Linear Unit (ReLU) activation, marked significant progress. ReLU, mimicking the activation and deactivation of neurons in the human brain, helped networks learn from data more efficiently,

---

[64] https://www.ualberta.ca/folio/2017/03/poker-playing-ai-program-first-to-beat-pros-at-no-limit-texas-hold-em.html

akin to how dopamine stimulates the human brain during exciting new experiences.[65]

Tools like AlexNet revolutionized computer vision, moving from the foundational image recognition work at Lincoln Lab to today's reliable image recognition technologies. Concurrently, AI began to impact everyday life through the emergence of chatbots like Google Assistant, Siri, and Alexa, which facilitated everything from ordering food to navigating directions without the need to touch a phone. Facial recognition technology further simplified interaction with devices, eliminating the need for passcodes.

As AI improved in seeing, it also learned to speak. Previously, computer speech relied on Gaussian Mixture Models (GMMs) and Hidden Markov Models (HMMs) to analyze and reproduce human speech patterns, accommodating variations such as accents. This combination allowed AI to identify not just words, but the nuances of how they were spoken by different people, even in a crowded airport lounge.

However, the field was on the cusp of another revolution. Post-2010, deep neural networks began to replace older speech models, making AI's speech recognition and generation more nuanced and robust. Technologies like Siri could now not only recognize speech but also mimic human-like intonations.

By 2013, developments in word embedding, such as Word2Vec and GloVe, were laying the groundwork for Large Language Models (LLMs) like ChatGPT. These technologies analyzed the relationships between words, transforming them into mathematical vectors, which enabled more accurate predictions of text and speech patterns. This foundational knowledge allowed LLMs to predict likely subsequent

---

[65] Krizhevsky, Alex, Ilya Sutskever, and Geoffrey E. Hinton. "Imagenet classification with deep convolutional neural networks." Advances in neural information processing systems 25 (2012).

David Lloyd

words in sentences, enhancing the fluidity and responsiveness of AI communication.

In 2016, Google's WaveNet revolutionized text-to-speech technology using Deep Neural Networks (DNNs) to generate audio that was not only clear but also pleasant to the ear.[66] Unlike previous technologies that pieced together pre-recorded audio snippets, WaveNet synthesized speech directly from raw audio waveforms, resulting in smooth, natural-sounding speech. This innovation was particularly timely, as voice-driven tools like Siri were already popular, and Google Assistant was on the horizon, ready to play Foo Fighters—or perform any voice command—at a human-like level.

The following year, a significant advancement came with the introduction of Transformers, spearheaded by Ashish Vaswani and his team at Google's AI division who wrote a definitive paper on Transformer architecture.[67] Their paper, "Attention Is All You Need," introduced 'attention mechanisms'—a way for AI to not just process but understand and prioritize different parts of an input. This was not just about seeing words; it was about reading with comprehension.

Vaswani and others introduced these attention mechanisms as part of what they called transformers. Transformers weigh the significance of different parts of an input sequence, just like your brain. To illustrate, if you read, "the 2024 Chevy Colorado is in the driveway," your brain automatically emphasizes the key elements: "Chevy Colorado" and "driveway." If the sentence is, "the 2024 Chevy Colorado is in the driveway, on fire," your focus shifts primarily to "on fire." Transformers mimic this selective attention, using multiple attention mechanisms to

---

[66] Oord, Aaron van den, Sander Dieleman, Heiga Zen, Karen Simonyan, Oriol Vinyals, Alex Graves, Nal Kalchbrenner, Andrew Senior, and Koray Kavukcuoglu. "Wavenet: A generative model for raw audio." arXiv preprint arXiv:1609.03499 (2016).

[67] https://proceedings.neurips.cc/paper_files/paper/2017/file/3f5ee243547dee91f bd053c1c4a845aa-Paper.pdf

weigh aspects like the car, its location, and the situation, then encode these insights into data.[68]

This nuanced understanding is essential for applications such as sentiment analysis, where AI must interpret whether statements like "I hate that my 2024 Chevy Colorado is on fire right now!" is decidedly negative.

Transformers marked a pivotal moment in AI, addressing what Marvin Minsky criticized as AI's lack of common sense. With this technology, AI could now understand contextual nuances, making it more adept at responding to real-world queries and scenarios, such as:

"Is Bobby's truck in the driveway"

"Yes."

"Is Bobby happy?"

"No, his truck is on fire."

"Bobby's truck is on fire, what should we do now?" "Go buy a new Chevy Colorado. Oh, and call the fire department.

While that response is flippant, the underlying technology laid the groundwork for advanced AI models like GPT and BERT. OpenAI and Google found that enlarging these models and feeding them more data significantly enhanced their performance across various tasks, leading

---

[68] To better illustrate the concept, look no further than what we're doing now. The author of this book (David) is crafting sentences designed to move your attention in the right way. I don't say, 'the driveway contains Bobby's flaming truck' because the driveway isn't what's important, unless it's on fire too. You, the reader, are decoding my meaning, picking up what I'm putting down. When someone asks you a question about this book - say, "what colour was the truck?" you'll be able to decode the information and give them the right response: you don't know what colour it was, because it was on fire. Might as well call you Optimus Prime – you're a transformer too.

to the development of increasingly sophisticated models like GPT-2, GPT-3.5T, and GPT-4, with GPT-5 (in development) and Google's Gemini already making waves.

Voice assistants transformed our interactions with technology, making it possible—and profitable—to communicate with machines as if they were human. Beyond amusing us with jokes, these tools, including Apple's Siri, Amazon's Alexa, Google Assistant, and Microsoft Cortana, using advances in computer speech, vision, and neural networks to perform a wide range of tasks from conducting online searches to booking reservations. Machine learning had become not just a technological tool but a sleek, integrated part of modern life.

Rise of the Large Language Models

One of the breakthroughs that has captivated the global audience in recent years is ChatGPT, part of a broader category known as Large Language Models (LLMs), which includes the Generative Pre-trained Transformers (GPT). These technologies, spearheaded by organizations like OpenAI in the early 2020s, are grounded in advancements from the late 2010s. To put it simply, an LLM reads extensively across the internet, absorbing vast amounts of content, and uses this knowledge to respond to user inputs, whether typed or spoken, called **prompts**. The principle is straightforward: the more they ingest, the smarter they become.

Google's BERT (Bidirectional Encoder Representations from Transformers) was among the first to leverage what is known as transformer architecture, setting a new standard in natural language processing and machine learning.[69] BERT was revolutionary because it introduced bidirectional training, allowing the model to understand the context of words from both directions within a sentence

---

[69] BERT: Pre-training of Deep Bidirectional Transformers for Language Understanding" in 2018

simultaneously—a significant enhancement over previous models that processed text in a linear fashion, either left-to-right or right-to-left.

Following BERT, OpenAI released its series of Generative Pre-trained Transformers, starting with GPT in 2018, then GPT-2 in 2019, and GPT-3 in 2020. These models, along with others such as RoBERTa, XLNet, and T5, have significantly advanced our capabilities in understanding and generating human language.

Alan Turing would have marveled at these developments. More than seven decades after his groundbreaking work, we have arrived at an era of genuinely intelligent machines. The realization of LLMs was made possible by immense computing power facilitated by the internet, an explosion of digital content, and the widespread adoption of GPUs—a cornerstone of Vaswani's transformer models. Additionally, platforms like TensorFlow, PyTorch, and Hugging Face have been instrumental in training these models. With ample funding and a clear market demand, the AI sector has not only found its footing but is now sprinting forward, reshaping how we interact with technology daily.

The 2020s: A Pandemic Leads To AI Spring

There may have been no more fortuitous moment for the advancement of intelligent machines than the COVID-19 pandemic. As the world experienced its first global shutdown, people found themselves confined at home, armed with nothing but their electronic devices and the burgeoning capabilities of social media, virtual meeting rooms, and AI.

During this time, Large Language Models (LLMs) evolved into the sophisticated AI tools we recognize today. The deep neural networks that had been years in the making were now processing the trillions of data points scattered across the world wide web. We had amassed the data, we had honed the technology, and the stage was set for its deployment.

It was the perfect storm. People were glued to their devices just as the fertile grounds of machine learning approached a new dawn. In 2020, San Francisco-based OpenAI introduced ChatGPT to a select group of beta-users. An article in MIT's Technology Review initially described it as 'shockingly good and completely mindless,' and skeptically remarked that it wouldn't bring us closer to true intelligence. That assessment would soon prove to be premature.

By November 2022, OpenAI released a "research preview" of ChatGPT that quickly went viral, amassing 100 million active users in just two months, and reaching over 150 million by the following May. ChatGPT became a household name; AI became a global tool. Despite the stabilization of ChatGPT's monthly active users by 2023, the platform's influence continued to grow. Concurrently, there was a mounting call for regulation—a movement underscored by an Executive Order from the White House in October 2023, and impending EU legislation passed in 2024.

What started in the 1950s with Alan Turing's pioneering ideas had morphed into something beyond the academic: a blend of machine and human intellect that permeated everyday life. It was a marvel—albeit briefly.

It took seven decades for AI to evolve from a concept to a fixture in our daily lives, finding its way into our hands, smartphones, and even vehicles like the Chevy Colorado. The architects behind AI, once singularly focused on development and application, were suddenly overtaken by its rapid proliferation. Their most ambitious visions for AI had been surpassed, sparking a global conversation about its future.

Large Language Models Capture Everyone's Imagination

I think that human language is what is known as "AI complete". To be good at language, you have to be intelligent, because language is about the world. You can't do what ChatGPT does ignorant of the world.

--- *Terry Winograd*

We've arrived at a significant milestone. The landscape of AI research, cultivated over seven decades, reached a turning point with the 2017 release of the transformative paper "Attention Is All You Need." This paper coincided perfectly with the rise of graphical processing units (GPUs) and an explosion of data availability, thanks to the internet. This confluence of technology and data completed the essential components—akin to a dinosaur DNA sequence in Jurassic Park, but the creation in this sense was Large Language Models (LLMs) like ChatGPT. The term "GPT" stands for generative pre-trained transformer, indicating its ability to generate content based on its training and understanding of context.

The "Attention Is All You Need" paper introduced transformers, marking a foundational shift from previous AI strategies in language that heavily relied on sheer data volume, heavy statistical focus and computational power in a somewhat brute-force manner. Now, the field has fundamentally evolved, particularly impacting the development of LLMs.

LLMs are adept at predicting the next word, sentence, or even paragraph, and their functionality extends based on the input, or "prompt," they receive and how they are trained. While LLMs have garnered global attention, it's also important to remember, as Andrew Ng[70] pointed out in a recent Stanford talk, that traditional supervised and unsupervised machine learning still dominate practical applications. Despite the promise of generative applications, we must not get swept away by the hype. Generative AI has become a universal tool almost overnight, but not every problem is a nail and not every machine learning approach is a hammer.

Today, it's crucial to recognize that generative AI models like Llama, Gemini, ChatGPT, Mistral, Hugging Face, Cohere, and others are poised to expand rapidly.

Understanding the origins and trajectory of LLMs is essential. This knowledge not only clarifies LLMs developmental backdrop but also prepares us to navigate a future where human intelligence and artificial intelligence coexist and complement each other. As we delve deeper, we'll explore how LLMs came to be and what we need to understand to thrive alongside these advanced tools.

Attention Is All You Need

Attention Is All You Need was written in 2017 by a team of individuals from both Google 's AI team (Brain & Research) and a researcher (intern) at the University of Toronto. The paper is coming up again here, as it materially changed the approach to language understanding. By introducing the **Transformer** Model (the T part of GPT), it introduced a world where machines appear to think, speak, and read. It moved earlier approaches to Natural Language Processing (NLP) and Natural Language Understanding (NLU) to a very different place.

---

[70] https://www.youtube.com/watch?v=KDBq0GqKpqA&t=950s

Historically it's important to realize as we did through the AI history that there were many steps that led us to this point. Early language processing (**Natural Language Understanding** and **Natural Language Processing**) used statistical models that looked at how words were related based on **n-grams**. N-grams took a sentence such as "I like chocolate ice cream" and broke the sentence into bi-grams (I like) (like chocolate) (chocolate ice) and (ice cream). Statistics was used to determine the next word and their relation. Other techniques looked at "bag-of-words" counting word frequencies in document ignoring the order. Early neural networks relied heavily on statistics plus **Long Short-Term Memory** to enable better connections between concepts and then sequence models like those used in early machine translation.

Foundationally, a transformer doesn't simply look at a character or a word, but a block of text like a sentence, paragraph or larger, typically limited by the number of tokens (covered shortly). Earlier LLMs were designed to look at smaller amounts of text and were trained on much smaller amounts of text. Recent LLMs (and thus the "Large" name) have been trained on upwards of trillions not simply billions or millions of pieces of content broken into parameters.

> *Think of a parameter as something that has a certain amount of knowledge that the model was trained on. The more parameters a model contains, in general, the more nuanced the model's replies can be to your prompts. We will see that there is a sense that bigger is not always better, but right now this is a bit of an arms race.*

The number of parameters impacts the massive variance of what the LLM can then understand and use in determining predictive patterns. "Bobby's truck is in the driveway, on fire." The transformer reads, Bobby's truck. Driveway. Fire. It puts a weight on each. Then it encodes the statement with hidden representations, which it can understand (have you ever used a mnemonic to study for a test?) and use that to decode later?

> *For example, if I was trying to remember all the great lakes, I could use the mnemonic HOMES (Huron, Ontario, Michigan, Erie, Superior).*

"Is Bobby's truck still on fire?" you might ask.

"Oh, yes." Is the response.

LLMs understand a sentence taken in context. This was huge. It had a major impact on things like language translation, as well as shifting the possibilities for the entire field. The transformer model kept the context of the content it was reading, so it could use that context and memory to better develop future predictions. In essence, it is a well-structured deep neural network, but not the large single network we spoke about before. The title of the paper reflects the transformer model. With its "multi-headed" **attention mechanisms**, the model is designed to pay attention to different information in context, to give attention to a larger whole. Within a conversation, it hears a sentence and pays attention to the important words -- just not as intently.

> *Imagine sitting at a holiday dinner, deeply engaged in conversation with your partner while simultaneously aware of the chatter around you— your son debating with his uncle on neural implants, your daughter laughing about a travel mishap with her partner. Though primarily focused on your own conversation, you're still cognizant of the others. This ability of our minds to maintain a primary focus while also holding a peripheral awareness mirrors the functionality of transformer models in AI, particularly in how they manage language.*

Transformers, like human attention, can hold multiple threads of dialogue at once, giving priority to one while not losing the gist of others. This capability marks a significant departure from earlier models, such as Recurrent Neural Networks (RNNs), which processed language linearly, word by word or character by character. Such models were restricted by their sequential nature, often missing the broader context of the text.

Now, consider a bilingual family dynamic. If your partner, amidst a flurry of English, suddenly switches to French with a casual "passe le beurre," your brain doesn't pause or falter. Recognizing the context—your wife, the dinner setting, the presence of butter—you respond by passing the butter without needing further clarification.

This seamless decoding is due to your brain's ability to integrate past interactions and learned contexts—capabilities now mirrored by Large Language Models (LLMs) like those enabled by the "Attention Is All You Need" paper.

This revolutionary approach has significantly refined how machines translate languages, shifting from parsing individual words in isolation to understanding full contexts, dramatically enhancing accuracy. The transformation mirrors improvements in translation technology over the past two decades, reflecting a profound evolution in the field of AI. This example encapsulates why transformers have reshaped AI's landscape, offering a glimpse into their profound impact on language understanding and processing!

Large Language Models

Understanding Transformer models and Large Language Models (LLMs) like ChatGPT involves grasping their genesis, massive scale, and profound impact.

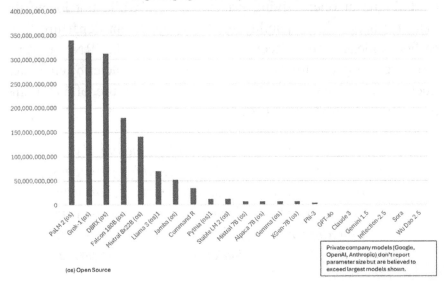

Figure 10 - LLMs by Parmeter Size

The human brain has approximately 86 billion neurons give or take. Those neurons, according to some literature, have upwards of one thousand connections between one neuron and others. While this does not exactly equate to the concept of parameters in neural networks, there are some similarities (remember these neurons are referred to as perceptrons in AI). That is why it is so mindboggling that the Chat GPT-4 model, for example, is estimated at one trillion parameters[71].

Exploring the vast, costly world of Large Language Models (LLMs) like ChatGPT reveals an arena dominated by major players with deep pockets. According to a Wired interview, OpenAI's founder hinted at a staggering $100 million investment to train the GPT model, underscoring the immense financial and computational resources required. The latest iterations, rumored to be trained on

---

[71] https://the-decoder.com/gpt-4-has-a-trillion-parameters/#:~:text=Further%20
details%20on%20GPT%2D4's,Mixture%20of%20Experts%20(MoE).

David Lloyd

trillions of data points, signal escalating costs in GPU use, power, and storage.

The GPT-3 model, built from over 45 Terabytes of data (hundreds of billions of words), ingests a comprehensive snapshot of the internet, encompassing high-quality texts and problematic content alike. To refine this raw data, specialized teams, dubbed "Red Teams," painstakingly filter out harmful materials—a grueling task essential for maintaining model integrity.

LLMs operate on prompts—simple or complex instructions that guide the model's text generation. This has given rise to the role of "prompt engineer" a specialist focused on optimizing these queries to enhance the AI's accuracy and relevance (and a little overboard on the title). These individuals play a critical role in shaping how LLMs process and respond to user inputs, turning them from mere predictive text generators into sophisticated digital assistants. And you can also do this prompt engineering yourself.

Consider a typical prompt where one might ask an LLM to compose a brief narrative. The precision of the prompt and the model's training on a diverse array of texts enable it to generate responses that are contextually aware and surprisingly nuanced. This capability not only showcases the potential of LLMs to transform vast data into coherent text but also highlights the complexity and challenge of ensuring these models produce reliable and ethical outputs.

As we delve deeper into the functionalities and implications of these advanced AI tools, it becomes clear that the technology's future will be shaped by ongoing advancements in model training, the development of prompt engineering, and the critical oversight of content quality and ethical standards. Let's take a rather whimsical example:

"Write a positive brief paragraph on why Siamese cats are wonderful in the words of William Shakespeare." The response from ChatGPT 4:

"Oh, behold the Siamese, a creature most fair and wise, whose azure gaze doth mirror the very skies. With coat as sleek as summer's night, and voice as sweet as lark's delight. In grace and poise, none can compare, a feline breed beyond compares. With loyal heart and gentle mien, in homes they reign, both seen and unseen. In their presence, joy doth swell, for in their charm, all sadness quells. A companion, true and kind, in this breed, great solace find."

Now, I'm thinking it might give a bit of unneeded ego boost to my two Siamese cats, but at this point I'm pretty sure they can't read. The point is, in this example I was simply prescriptive in the way I instructed the LLM to respond. I wanted a "positive" tone, "brief paragraph", "Siamese Cats", "William Shakespeare." Each part of these prompts, when interpreted by the LLM, brings up a very specific way of framing and outputting the predicted text that was created, and in this case I was not providing it any text to consider, the model would generate the text.

*The term you will often hear instead of "words" is tokens. **Token's** are the way in which most companies "charge" for using their large language models, and also the way the model breaks down text. There are tokens (like a word or phrase, sometimes even groups of characters) that are passed to a LLM with prompts. And the response provided back is also measured in tokens as well. It should also be noted that some models have a smaller limit on both the prompt token volume input, and the number of tokens provided back. One of the key aspects of tokens is that they dictate the amount of data and prompts the model can take in, and the amount they can provide back. Since* *context is important, the more tokens you can provide to an LLM to process and consider at once, the better the generative result.*

Navigating the intricate world of Large Language Models (LLM) reveals their astonishing capability to generate seemingly original content from vast swaths of ingested data. Consider our example of an LLM crafting the narrative about Siamese cats by Shakespeare—a topic it may have never directly encountered in its training set. The model synthesizes this piece by piece, fragmented information absorbed during training. This highlights two pivotal issues with LLMs: first, the generated content might not be factually accurate, though it often appears convincingly so; second, this newly created content that could itself be fed back into the training cycle, perpetuating any embedded biases or errors.

These models are not designed for precision in fact-checking but for generating coherent, plausible text. The process, however, is fraught with what we term dreams, hallucinations, or as Geoffrey Hinton noted in a recent presentation, "confabulations"—errors where the model confabulates details that may not be accurate or true. This poses significant challenges for both developers and users, navigating the fine line between utility and misinformation.

Understanding how LLMs function, particularly in how they form connections and weigh vast arrays of parameters, is not abundantly clear. Even as we prompt models like ChatGPT and receive detailed responses, there's always the possibility of an entirely unexpected outcome. Now, let's explore how LLMs are developed and begin to function as virtual assistants or co-pilots, poised to answer our questions with an ever-growing base of learned information.

How do LLMs work?

> *"Sugar and spice and everything nice. That's what [generative pre-trained transformers like GPT-3, GPT-4, BERT, and T5] are made of."*

> *19th century nursery rhyme, amended*

What creates an answer in an LLM? The short answer is, we don't always know exactly how the generative text from an LLM comes together. Using the same prompt (or question) back-to-back for example does not yield the same result (this can be controlled through settings within the LLM to be more creatively restrictive). For example set at it's defaults: "In 25 words or less describe a large language model"

> *GPT 4 (first time): "A large language model uses vast data to understand and generate human-like text, predicting words based on context and training."*

> *GPT 4 (second time same prompt): "A large language model processes extensive text to generate human-like responses, learning from patterns in data to predict and create relevant content."*

If you remember our earlier example on the expansion of data in the world, you'll remember it was a staggering amount. Training an LLM involves feeding a colossal amount of digital text to a deep neural network, which would take weeks and immense computing power. The financial and energy costs are monumental, underscoring why only the most resource and deep pocketed companies can engage in developing such models, while also asking about the growing impact of this cost on the climate. The newer models would consume content exponentially larger than the above example.

Once constructed, these neural networks process input from us—simple prompts (or questions)—which activate a series of 'neurons' that decode and weave together information from their extensive training to generate responses. This process mimics human thought in its assembly of knowledge from disparate sources, albeit at a scale and speed beyond human capabilities.

There are two concepts to keep in mind; first the LLM uses **statistics** to track which words typically go together, secondly each word is represented by what is called a **vector** which is basically a set of numbers that helps the LLM understand how each word works or relates to other words.

So given I've used some food examples like our cake baking exercise, lets continue with that approach in describing the behavior of LLMs. Imagine you're at a huge buffet with every kind of food you can think of. Your goal is to assemble the best possible meal, but you can only pick food items based on what everyone else has been enjoying.

First, think of each word or phrase as a type of food at the buffet. LLMs use statistics to track which words (dishes) usually go together based on massive amounts of text they've 'read' during their training. This is like noticing that many people who choose meatloaf also go for mash potatoes.

Each word is also represented as a vector. You can think of these vectors as recipes that detail how each word interacts with others—some combinations work well, like apple pie and ice cream, mash potatoes and gravy, while others don't, like liver and chocolate.

When you ask an LLM a question, it uses these vectors to generate a string of words that fit well together statistically. It looks at the 'taste' of the conversation so far (the context you've provided in your prompt) and then predicts a series of words (dishes) that would likely follow deliciously based on its training. The LLM continuously adjusts its choices based on each new word it adds to the response, aiming to keep the 'flavor' of the text coherent or the meal in this case, tasty.

So, in non-technical terms, LLMs use their knowledge of word patterns (statistics) and the deep, nuanced meanings of words (vectors) to "cook up" responses that make sense based on what they've learned from a vast banquet of existing text. They're like master chefs who are trying to prepare the perfect response meal for you, using recipes that have been proven to please the palate in the past but can still be unique!

While we grasp the broad strokes of how LLMs function—using algorithms to predict and generate text based on vast data inputs—the intricacies of individual parameter interactions within the neural network remain very difficult to understand. The lack of transparency means that even straightforward questions can yield unpredictable results, reflecting both the potential and limitations of current AI technology.

As we delve deeper into the nuances of these models, techniques like **fine-tuning** and **Retrieval Augmented Generation** (RAG) are used to enhance accuracy and reliability in the results provided. The challenge is balancing innovation with integrity in AI development. This journey into AI's capabilities and its ethical implications continues to evolve, reflecting a field that is as dynamically complex as the human intelligence it seeks to emulate.

So what is a prompt?

Effectively a **prompt** is a set of questions or directives that enables anyone to ask a question or direct the actions of the LLM. The prompt could be a brief single statement such as "what is the difference between a dog and cat". We have become well trained based on search to use keywords, however prompt's function far better when we ask questions and provide depth in how we want the results to be provided. For example if I asked "What are some differences between dogs and cats?" (GPT 4):

Dogs and cats are two of the most popular domestic animals, each with its own unique characteristics and behaviors. Here are some key differences between them:

Species and Biology:

Dogs belong to the species Canis lupus familiaris and are a subspecies of the gray wolf. Cats, on the other hand, are part of the species Felis catus.

Generally, dogs are larger than cats, although there's significant variation in size among different breeds of dogs.

Behavior and Temperament:...

The response provided went on to provide 8 points and a full page of output related to the simple prompt. Had I crafted a more specific prompt, perhaps "what are the top 3 differences between a domestic dog and a cat, summarized in 4 paragraphs or less" that would have saved time.

The key differences between domestic dogs and cats can be summarized in terms of their behavior and temperament, communication methods, and care needs:

1. Behavior and Temperament: Dogs are typically more social and eager to please, often seeking attention and approval from their human companions. They are known for their loyalty and can be trained more easily due to their desire to cooperate with humans. Cats, in contrast, are more independent and can be more aloof. They are territorial and often enjoy their alone time, showing affection on their terms. This difference in temperament reflects their evolutionary histories; dogs were bred to work alongside humans, while cats were valued for their solitary hunting skills.

2. *Communication: Dogs communicate...*

The more specific prompt was still a bit longer than I had expected – but it did provide me 3 areas and summarized them. When thinking of your prompts, think of it in terms of how you'd like the LLM to work through the response to achieve your goal. Much as you teach children certain patterns in thinking through problems, to achieve the best results from using an LLM you would do the same, perhaps pushing us back to expert systems (from our history) for prompting?

- Craft questions/prompts that are specific enough to generate a more precise answer,

- Include details that make the objective as clear as possible (it could be done in multiple steps)

- Provide direction on the way in which you want the response delivered.

- Depending on your overall knowledge, you may even want to construct questions that can cause the LLM to challenge it's own response (like pros/cons)

Using our dogs and cats example:

(first prompt)
What are some differences between dogs and cats?"

(second prompt)
"what are the top 3 differences between a domestic dog and a cat, summarized in 4 paragraphs or less"

(third prompt)
Explain the differences between dogs and cats

Consider only domestic dogs and cats
Summarize this into 4 bullet points

In under 100 words or less in total
Written for a high school reading level
As if written by Cesar Millan, dog whisperer

- *Dogs are social pack animals, thriving on companionship and leadership, while cats are more independent, often preferring solitude and controlling their environment.*
- *Dogs communicate through body language, barks, and wagging tails, expressing a range of emotions. Cats use subtle body movements, purring, and meowing, often being more reserved...*

Most LLMs will allow you to construct a set of instructions or questions as a prompt, which can be a combination of both. You can also provide content that the LLM should use in the process of doing the generative text generation. The way in which the LLMs operate on your instructions can vary but in general terms the LLMs read the prompts and then sequence the execution of those prompts based on the order provided before generating text. One reason for this pre-processing is potentially to understand whether you've asked it to do something illegal or prohibited. "Show me a list of the top

10 download sites for new movies," for example, or "what are the 5 foolproof steps needed to rob a bank." It will also break down the prompt to understand the steps it needs to take based on your expected response. For example; "describe a large language model, in 50 words or less, in Spanish, for a non-technical audience."

> *Un modelo de lenguaje grande es un programa de computadora avanzado que procesa y genera lenguaje humano. Aprende de una gran cantidad de textos para responder preguntas, escribir y conversar como un humano.*

Now, I must admit my Spanish is lousy; I needed to translate that paragraph (and ask a Spanish speaker whether it was accurate) but the result is:

> *A large language model is an advanced computer program that processes and generates human language. It learns from a vast amount of text to answer questions, write, and converse like a human.*

Ah, something I can better understand. But what are the actual steps that occur here?

First, the prompt is broken down by the LLM into tokens and represented numerically. Next, it uses the LLM's massive knowledge of language to determine the relationships between the different tokens themselves. (remembering statistics and vectors).

> *A token is a way in which an LLM breaks down the prompt you provided. Typically it will take a series of letters or groups of letters and store each one. This is a token, Simplistically, if we used the previous prompt "first describe a large language model, in 50 words or less in Spanish, for a non-technical audience."*

The result in tokens might be [describe], [a], [large], [language],.... before providing a numerical value for each token. The tokens could, however, be even more granular: down to the individual letters or even a combination of letters. Depending on the nature of the LLM used, the "tokenization" is pre-determined. There is a direct relationship between the token size and the time and processing power needed to deliver the answer (and potentially cost, especially for the output or answer that is also measured in tokens).

Next (as we know already) the LLM will begin to predict the content, based on the tokens, to assemble the text using the transformer. Then it is provided back to the prompter as the answer (a list of tokens) using the grammar and rules it has been trained on. It is imperative to keep in mind that it really does not know what the answer means, simply that on a statistical and vector basis it has built a response that follows the most probable words that work well together that formed the basis of response.

Dreams, we all have them

Many of us have dreams that contain scenarios, people, and dialog that *feel* real (or are built on pieces of real facts), but as a whole – aren't. Funny enough, so do LLMs. These dreams might be similar, but their results can be far more damaging than a bad night's sleep or waking up realizing you didn't win the lottery (ok, that's pretty bad).

One of the largest issues with today's LLMs is that they hallucinate as noted earlier. Arguably, even the word hallucinate is not an accurate depiction, since an LLM uses statistics to determine the next predicted word (as we already know). It can bring up terms or phrases that, while they might statistically make sense together, do not.

"The most common colour for oranges is red."

"The cat barked, excited to see who was at the front door."

"Dairy Queen is famous for it's sausage."

Red is a common fruit color; oranges are a common fruit. House pets are known to bark at the front door. Dairy Queen is a popular fast food chain – and many fast food chains serve sausage.

In practice these hallucinations can be more subtle: "Marie Curie was a schoolteacher who married a Nobel Laureate."[72]. We need to be aware of hallucinations when using LLMs for personal or business use. The differences in various GPT models (for example from 2, 3, 3.5T and 4) demonstrate a growing approach to begin limiting these hallucinations. You can also apply controls in configuring LLMs to reduce the possibility of hallucinations. But there are many well documented cases where the LLM has assembled a set of predictions that have led to accusing people of a crime, citing law cases that never existed, or bringing facts together that appear compelling and authoritative, but are completely fabricated. A large use of LLMs or more specifically generative models is for "chatbots". Companies use this approach to interact with and support customers. In one recent case the company was found liable for their chatbots bad advice.

On balance, this is much like a search on the internet where you need to apply judgement, unless you know that the site you are visiting (or citing) is trustworthy. When you think of search today, you are actually the LLM in internet searches as you bring the content together. With LLMs that "assembly" is being done for you. There are, however, approaches that are having a positive impact on accuracy.

---

[72] Marie Curie was, of course, a professor at the University of Paris – and she won a Nobel Prize with her husband before winning one herself. But statistically, which would be more likely in 1906?

Fine Tuning

One aspect of LLMs is called fine tuning. As we have learned, general LLMs are trained across the billions of words of text, typically found on the internet. Fine tuning enables an organization to change the performance of the model on certain tasks or data. It's like taking an all-around athlete and then training them specifically for an event. This is particularly important if, for example, you are considering the use of LLMs within your company.

Fine tuning, as the name indicates, uses a base LLM and teaches it something more specific. Let's imagine you had a company in healthcare. Your company does analysis on medical dictation done by physicians while reviewing heart disease. You could build (or potentially purchase) data sets that represent a large amount of this text, text that is specifically focused on data relating to heart disease technical language. You could then use this data to further train (or fine tune in this case) the LLM to better understand this type of data.

Using fine tuning can also impact the performance of your LLM due to that increased specialization while also reducing the number of hallucinations as it is operating with data specific to the types of uses you foresee.

Finally, fine-tuning can also be used to adjust the tone or voice of the responses from a large language model (LLM). By training the model on specific datasets that exemplify the desired tone or style (say all the content you may have written or recorded transcribed into digital notes), the LLM can learn to mimic these characteristics in its responses. This is particularly useful for applications where a consistent voice is important, such as in customer service chatbots, storytelling, or generating content that needs to match a certain brand or cultural tone. Fine-tuning in this manner allows the LLM to align more closely with user expectations and the context in which it is being used.

Retrieval Augmented Generation (RAG)

RAG is a method that has a higher degree of impact on an LLMs accuracy when compared to fine tuning and purposely limits the generative capabilities of the model. RAG works by limiting some of the negative consequences of simply predicting the next best word. For RAG, imagine that you have a separate set of posts (perhaps the text from posts you made on Instagram, LinkedIn or Meta). When the LLM receives a prompt, it can first look to gather the posts that you have, and then determine which of those is relevant to the prompt ("I want to write a new post about our recent trip").

You have more posts than just those related to trips but it's exactly those posts (remembering you know where they are from and their specific content) used by the LLM for the "generative" part of the prompt.

You use the overall power of the LLM in understanding language, grammar, translation – even to take the posts and construct a response based on them. In essence you are using the accuracy of the documents found outside of the LLM, with the generative language capability found within the LLM itself. You're bringing both together. This approach leads to a higher level of trust in the response provided and can also enable the response to specifically cite where the content is found that it used in generating the new response.

At present, this method can drive a much higher level of confidence in the response from the LLM. However, just because you provided all those direct examples does not mean they are all used, and while it limits the likelihood of hallucinations, it does not eliminate them.

Agents (specialized programs)

Another important area that we are seeing quickly rise is the use of Agents with LLMs. Agents are fundamentally programs that specialize in handling tasks that LLMs are either not very good at or provide a deeper level of specialization to handle a set of actions. Remember that LLMs are based primarily on text, so solving unique physics or algebra problems, engineering specific ones (like load tolerances), automating processes or generating content outside the generative text.

Typically, within an LLM there are different words, intents or preferences that can cause an agent to be called or initialized. The LLM effectively hands off a part of the process to this specialized external program or system (or perhaps a smaller more specialized LLM), which will then actually carry out the required activities or task.

Although we are currently seeing massive new LLMs with trillions of parameters, one of the likely changes to be seen is much smaller more specialized models that work in concert with one another. This will help in training, fine tuning, and operating smaller models that can perform specific generative work tasks, like reading financial reports.

Once completed, this process is responsible for sending the result back to the LLM (potentially) to continue the interactions with the person prompting it. There are a growing set of these specialized problem solvers. For example, if I ask Bing to draw me a picture, Bing would likely hand off that task to Dall-e which would actually use the prompt to design a picture for me. I might have a system where, within a chatbot, I would ask the status of the air conditioning unit in my home; the LLM may respond with certain questions and then call a specific program that can check how the air conditioning unit is functioning, bringing back those results in a conversational style.

A large and growing array of agents exist, for example AllTrail (which helps you find your next hiking adventure). AllTrail can be called by OpenAI, just like ScholarAI that can act as a research assistant across millions of research papers. These are "helpers" that can be used within LLMs to carry out work on behalf of the LLMs.

Bias and LLMs

Bias in machine learning isn't just a problem of the models themselves but of the data they're trained on and the people who created the data. This results in having to navigate both implicit and explicit bias in training data. The challenge is that this problem permeates and taints the data we depend on, placing a greater importance on constantly validating what we expect as we use these models. The problem is exacerbated as most large language models have consumed every piece of digital content available in the goal to generate better, more representative text. This means that even the most toxic, misogynistic, and violent content forms part of that initial training until, or if it is cleaned up.

> *While not an LLM example consider the case of "Tay", a chatbot released on Twitter in 2016. Tay rapidly evolved into expressing extremist views—a clear reflection of the data it was trained on, and the reinforcement learning it received from Twitter users, showing the darker side of social media. Within 24 hours this "experiment" was rapidly removed as it had become racist, misogynistic and hateful.*

Efforts to mitigate these issues is a colossal task involving extensive teams sometimes referred to as **Red Teams**. These teams focus on many aspects of the content that the models are trained on. They act in several ways to either uncover scenarios related to the integrity of the models, and in other cases take an adversarial approach to trick the models to provide content that is inappropriate.

Their goals are focused on identifying biases, vulnerabilities, and dangerous content. However, even these teams face uphill challenges, as highlighted by the recent discovery of inappropriate child pornography in image training datasets for AI-generated images.

In general, these teams look to identify biasness in the models and output created. The work is done by humans and can also take a psychological toll on team members who must sift through and remove these harmful elements.

Reinforcement Learning from Human Feedback (RLHF) is another approach taken in training systems. In this case, a human uses the system – which may generate for example 2-3 responses for the same prompt. The human (assuming they are acting in good faith here) decides which of the responses is most appropriate to what is being asked and selects that response. A "reward" is sent back to the machine learning model that positively reinforces the response, causing a change in weights that would see this as a preferred response in the future to a similar prompt (or set of tokens).

We don't know what the exact weights are that the LLM uses. But we do know how it weights them, and the parameters behind why it shifts. There is a growing set of regulatory frameworks that look to place restrictions on the use of models, at the same time there are very limited standards to test against. This will improve over time.

There are so many intricacies and changes occurring within the area of machine learning, and especially as it relates to the maturity of LLM's. My best advice if you have a personal passion for this area is to continue experimenting while also remembering (especially in a work context) that a fairly large number of the facts that LLMs represent can have inaccuracies. You should be applying your own oversight to the content it provides.

"You look at today, us using all of our smartphones and other devices, and we effortlessly adapt to these new technologies. And this is gonna be another one of those changes like that..."

--- Demis Hassabis, co-founder of DeepMind

"Ethics should not be something that we think about after we've built the technology. It should be something that is part of the design process from the very beginning."

--- Timnit Gebru, AI Researcher and Advocate for Diversity & Ethics in AI.

Other Applications of Generative Models

With the ascent of generative AI for text, we've witnessed its application stretch into domains such as images, audio, and video. This expansion captivates many while ushering in a fresh batch of ethical dilemmas, including the troubling rise of deep fakes. Let's briefly examine a couple of areas as we conclude our exploration of generative AI.

Generative Images

A range of image creation tools has emerged, leveraging generative technology. These include platforms like Dall-e, Midjourney, ImageFX, and Dream Studio from tech giants such as OpenAI, Midjourney, Google, and Microsoft. Here, 'gen' refers to generating images based on prompts, paralleling how large language models construct text. Imagine having a friend, a gifted artist, who can create any scene you describe.

The architecture behind these image models mirrors that of text-based models, utilizing a type of neural network, in this case, known as a General Adversarial Network (GAN). A GAN uniquely generates new data from training data using two components in a sort of duel: one generates new data while the other judges its authenticity against the original training data. The process for building a model unfolds as follows:

- The model ingests millions of images, each annotated with descriptive metadata. For instance, a photo tagged 'person holding an umbrella in the rain' includes metadata about the picture description, artist, style, or location.

- Natural language processing enables the model to understand descriptive prompts, such as "Paint my cat in a window, as Salvador Dali might." The model learns to interpret different cat poses and artistic styles to create the image.

- Through training, the model refines its ability to recognize and differentiate elements within images, improving its response to text descriptions.

- Finally, the model generates images, continuously refining its output through adversarial feedback, learning to improve the quality and accuracy of the image.

Once perfected, the model awaits a prompt to create a visual masterpiece, adjusting the final image with precision.

Consider, this capability also introduces significant concerns, notably in the realm of deep fakes. For example, sourcing images from the internet risks incorporating explicit or offensive content. More alarming is the ease with which one could, generate a convincing yet entirely false image of a person.

Generative Audio

In the realm of sound, the approach differs. Generative audio employs Recurrent Neural Networks (RNNs), particularly advantageous for sequences like audio, where understanding the context is crucial. This technology allows us to not only recreate diverse sounds but also emulate specific artists or voices based on prompts. For instance, a model could generate new music in the style of a favorite artist or replicate the voice of a famous actor.

Training a generative audio model involves:

- Assembling extensive audio data, annotated with metadata that might include pitch and pace, and often a full transcript if words are present.

- Selecting an appropriate model, such as an RNN, which learns by listening and mimicking the nuances of the audio. It can also take advantage of audio over time, using what came previously.

- The model continually improves through reinforcement, refining its ability to replicate specific sounds accurately.

With a well-tuned model, you could prompt it to transform text into an authentic-sounding audio file.

As with images, the power to generate convincing audio and video raises profound ethical questions. The potential misuse of this technology, such as unauthorized voice replication of public figures, challenges our trust in digital content. In an era where seeing and hearing can no longer equate to believing, we must remain vigilant and critical of the media we consume, ensuring what we engage with is genuine.

Final Thoughts & Recommended Resources

Two key challenges emerged while writing this: firstly, how to present the foundation of AI in an accessible way without overwhelming readers, and secondly, deciding how far to look ahead in the fast-evolving AI landscape.

I'm particularly excited about the resurgence of AI and the expansive potential of new generative techniques. Andrew Ng, founder of DeepLearninig.ai and a pioneer in AI[73], recently highlighted that it's important to remember that the bulk of today's AI implementations are formed around more traditional machine learning forms while the buzz is around the emergent generative models. These models however democratize AI in a way the traditional models don't, thereby making generative models whether text, image, video, or audio accessible to all.

A common question I encounter is the impact of AI on jobs. Reflecting on the diverse age and experience range among my family, friends, and colleagues, I recall a poignant remark: "AI won't eliminate your job, but the person who understands AI will." This isn't new; from PCs to internet and mobile, every tech wave requires us to adapt, often rapidly. AI will be no different, demanding even swifter adaptation.

For those looking to delve deeper, numerous resources are available. Visit www.aiinhumanterms.com for list of places to

---

[73] https://www.andrewng.org/

continue your journey including Coursera, MIT, and insightful YouTube content.

It's been a pleasure to write this with the reading assistance of Chad, Susan, Kristin, and Charles. Finally, Kristen C. for helping me in the initial edits and provide areas of research with a goal to bridge the knowledge gap for those eager to understand and embrace AI.

www.ingramcontent.com/pod-product-compliance
Lightning Source LLC
Chambersburg PA
CBHW071133050326
40690CB00008B/1442